UNDERSTANDING
ISO 9000
AND
IMPLEMENTING
THE
BASICS
TO
QUALITY

QUALITY AND RELIABILITY

A Series Edited by

EDWARD G. SCHILLING
Coordinating Editor
Center for Quality and Applied Statistics
Rochester Institute of Technology
Rochester, New York

RICHARD S. BINGHAM, JR.
Associate Editor for
Quality Management
Consultant
Brooksville, Florida

LARRY RABINOWITZ
Associate Editor for
Statistical Methods
College of William and Mary
Williamsburg, Virginia

THOMAS WITT
Associate Editor for
Statistical Quality Control
Rochester Institute of Technology
Rochester, New York

ADDITIONAL VOLUMES IN PREPARATION

UNDERSTANDING ISO 9000 AND IMPLEMENTING THE BASICS TO QUALITY

D. H. Stamatis
Central Michigan University
Mount Pleasant, Michigan, and
Contemporary Consultants
Southgate, Michigan

Marcel Dekker, Inc.   New York • Basel • Hong Kong

Library of Congress Cataloging-in-Publication Data

Stamatis, D. H.
    Understanding ISO 9000 and implementing the basics to quality / D. H.
Stamatis.
        p.   cm. — (Quality and reliability; 45)
    Includes bibliographical references and index.
    ISBN 0-8247-9656-X (hardcover: alk. paper)
    1. ISO 9000 Series Standards. 2. Quality control.   I. Title.   II.
Series.
TS156.6.S72   1995
658.5'62—dc20
                                                                95-32191
                                                                    CIP

The publisher offers discounts on this book when ordered in bulk quantities. For more information, write to Special Sales/Professional Marketing at the address below.

This book is printed on acid-free paper.

Marcel Dekker, Inc.
270 Madison Avenue, New York, New York 10016

Current printing (last digit):
10 9 8 7 6 5 4 3 2 1

PRINTED IN THE UNITED STATES OF AMERICA

In honor
of my father,
Charalambo

About the Series

The genesis of modern methods of quality and reliability will be found in a simple memo dated May 16, 1924, in which Walter A. Shewhart proposed the control chart for the analysis of inspection data. This led to a broadening of the concept of inspection from emphasis on detection and correction of defective material to control of quality through analysis and prevention of quality problems. Subsequent concern for product performance in the hands of the user stimulated development of the systems and techniques of reliability. Emphasis on the consumer as the ultimate judge of quality serves as the catalyst to bring about the integration of the methodology of quality with that of reliability. Thus, the innovations that came out of the control chart spawned a philosophy of control of quality and reliability that has come to include not only the methodology of the statistical sciences and engineering, but also the use of appropriate management methods together with various motivational procedures in a concerted effort dedicated to quality improvement.

This series is intended to provide a vehicle to foster interaction of the elements of the modern approach to quality, including statistical applications, quality and reliability engineering, management, and motivational aspects. It is a forum in which the subject matter of these various areas can be brought together to allow for effective integration of appropriate techniques. This will promote the true benefit of each, which can be achieved only through their interaction. In this sense, the whole of quality and reliability is greater than the sum of its parts, as each element augments the others.

The contributors to this series have been encouraged to discuss fundamental concepts as well as methodology, technology, and procedures at the leading edge of the discipline. Thus, new concepts are placed in proper perspective in these evolving disciplines. The series is intended for those in manufacturing, engineering, and marketing and management, as well as the consuming public, all of whom have an interest and stake in the improvement and maintenance of quality and reliability in the products and services that are the lifeblood of the economic system.

The modern approach to quality and reliability concerns excellence: excellence when the product is designed, excellence when the product is made, excellence as the product is used, and excellence throughout its lifetime. But excellence does not result without effort, and products and services of superior quality and reliability require an appropriate combination of statistical, engineering, management, and motivational effort. This effort can be directed for maximum benefit only in light of timely knowledge of approaches and methods that have been developed and are available in these areas of expertise. Within the volumes of this series, the reader will find the means to create, control, correct, and improve quality and reliability in ways that are cost effective, that enhance productivity, and that create a motivational atmosphere that is harmonious and constructive. It is dedicated to that end and to the readers whose study of quality and reliability will lead to greater understanding of their products, their processes, their workplaces, and themselves.

Edward G. Schilling

Preface

Since I was first introduced to the ISO international quality standards, I have devoted a tremendous amount of time to learning to understand them and applying that knowledge by training personnel and helping organizations obtain certification. My exposure to the international standards through the American Society for Quality Control (ASQC) led to my commitment to helping organizations—both manufacturing and nonmanufacturing—to standardize and improve their quality systems.

As a quality professional for over 28 years and as a consultant in the field of quality for the last 10 years, I have observed many quality systems, some with no definite mission, goal, or standardization. I have seen major companies demand their own systems or programs in their supplier base and define their own standards of customer service and satisfaction. I have seen the supplier base of major companies totally confused and ready to rebel because requirements differed from company to company and in some cases were contradictory.

The international standards system known as the ISO 9000 series offers a basic foundation in quality with the opportunity to build in that quality, in a more demanding role, throughout the organization. It offers a standardized base for all to work from, and at the same time it allows for individuality.

This book is about the ISO 9000 series. It addresses the evolution and rationale of the standards system, its structure, the interpretation of each of the elements, and its relationship to other standards. It outlines a detailed approach for the implementation of ISO in any organization, using project management. It addresses the issues important for optimization of the proposed implementation process and proposes a generic training curriculum that can be used in various organizations. The concerns of service, software, and environmental industries are examined, and ISO and other quality systems and programs are compared.

The book is written with two basic objectives in mind. First, it provides a reference for the ISO 9000 material. Thus, the target audience is viewed as anyone who is interested in or involved with quality issues. Second, it provides specific application methods and tools for the ISO implementation process that will result in attaining certification as well as keeping it. Thus, the target audience is also viewed as the individuals in any organization who are responsible for the implementation process.

In addition to these objectives, the book is designed to be used as a college text or reference at both the undergraduate and graduate levels. The book provides a scientific base of research for adult learning, motivation, the implementation phase, audits, and curriculum development.

Acknowledgments

In writing a book, many hours need to be devoted to writing, editing, and confirming thoughts, issues, and concerns with friends. This book is no exception. Many people have contributed to its development either directly or indirectly.

I want to thank my chief editor, my motivator, and the person who put up with me in my most irritating times. Without her, this book would not have been written. This person is my wife, Carla. I also want to thank my children, Christine, Cary, Stephen, and Timothy, for the patience they showed during my long writing days without complaining.

I want to thank the ASQC for giving me permission to use the ANSI/ASQC A8402-1194 vocabulary standard. I want to thank Mr. J. R. Roberts for providing current software information through the draft newsletters. I want to thank Mr. E. Mann from Transamerica Leasing and Mr. R. Williams from Positronix for sharing their thoughts on registration with me. I want to thank Mr. W. Harral and Mr. Doug Burg for introducing me on a formal basis to the concept and the application of ISO. I want to express my gratitude also to Mr. R. Peach for sharing with me some of the concerns in the implementation stage of the ISO.

Furthermore, I want to thank all the participants in my public seminars over the years, for I owe them a very strong "thank you" for offering suggestions and specific recommendations. I have tried to incorporate as many as possible. Without their comments and their input, this book would not be possible. Finally, I want to thank Mr. R. Munro and the Editors of the Quality and Reliability Series for their constant support and encouragement and for reviewing and offering suggestions for improvement.

D. H. Stamatis

Contents

1

A General Introduction to Quality Standards

In this chapter our aim is to address the issue of quality standards and to demonstrate that the concept is not new, by summarizing some of the historical facts on quality. We will show that standards and certification procedures have been around for a long time. However, the difference between the early period and our modern time is that the process to certification is now more sophisticated and can be replicated on a more consistent basis.

In addition, we will provide an overview introduction to the standards of the International Standards Organization (ISO) and discuss the need for such standards not only in the European Union, but also in the United States and world markets.

As defined by ISO 8492 (BS 4778), quality is *the totality of features and characteristics of a product or service that bear on its ability to satisfy stated or implied needs*. However, whose needs does the service or product address? Who are its customers? How do we define these needs? The questions are not easy to answer. In fact, the ISO has added seven footnotes to its definition, including: *in a contractual environment, needs are specified, whereas in other environments, implied needs should be identified and defined* and *needs can change with time.*

Within this definition we can identify ideas of fitness for purpose, value for money, reliability, customer satisfaction, environmental impact, versatility, compatibility with other products, maintainability, conformance to requirements, or other desired characteristics. These concepts of quality are not new, nor are they restricted to any age or culture.

In the laws of the kingdom of Eshunnana—about 2000 B.C.—we find requirements dealing with interest rates, type of investments, and penalties (Goetz, 1973). In Hammurabi's code—about 1730 B.C.—we find penalties for malfeasance (Meek, 1973).

In the days of the Egyptian pharaohs there was an extensively documented quality system relating to the burial of the nobility (Durant, 1954). This was known as the *Book of the Dead*. It described the manner in which the requisite rituals should be carried out and specified how the funerary goods to be buried with the deceased should be prepared. The purpose of this system was to ensure that the deceased enjoyed an afterlife that was at least comparable with his or her life on earth.

Achievement of the required standard was attested to by the application of the mark of the Superintendent of the Necropolis. In the case of Tutankhamen, we find what is probably the world's oldest and most famous quality failure. He was buried in a hurry, and the marks on two of the beds used in the embalming process show that the horizontal members were transposed—a situation that has parallels in modern industry.

The first Emperor of China, Qin Shi Huangdi, who was responsible for the vast, underground, terra-cotta army at Mount Li, decreed that all goods supplied for use in the imperial household should carry a mark that identified the maker so that if an item proved faulty he could be identified and punished (Durant, 1954). This was, indeed, a form of third-party certification.

In the Roman era we find for a first time that the external audit is instituted and specialists known as *Argenterii*—dealers in silver—were required to keep certain records (Corns, 1968). On the other hand, the Bible gives us the byword of quality systems: "An ounce of prevention is worth a pound of cure."

During the Byzantine Empire we find that every action was regulated by procedures that had to be followed to the letter. To enforce these procedures, the local governor had attached to his court retinue an official inspector, a *Logothete*, who was charged with the inspection of all workshops and operations performed in the district. If such an inspection disclosed an infraction of the rules, the *Logothete* would have the culprit brought to trial (Guerdan, 1956).

In large, stone European buildings, stones can be found that have the registered marks of the quality of goods produced by their members. The mark serves as a reminder that the Master Mason approved the work. Furthermore, the mark was a point of reference to the exact location and was useful for payment purposes (Allcock and Unsworth, 1991).

A similar situation existed with the merchant guilds. The products produced by their members were held to a much higher standard than everyone else's goods. In fact, it is said that the merchants who bought cloth that bore the mark of the Colchester guild rarely troubled to open the bales because the mark of the Colchester guild was so powerful that it guaranteed a certain level of quality. Given that mark, the quality of the product was expected (Allcock and Unsworth, 1991).

In 1140, a system of hallmarking was introduced to bear witness to the quality of gold and silver items. Except for changes in the duty mark, this has remained unchanged to the present day.

Defense and quality have always been close partners and two British literary figures were prominent in establishing quality requirements in the defense industry of their age.

Geoffery Chaucer, as part of a varied military and diplomatic career in the last half of the 1300s, was Surveyor of Supplies for the Royal Wardrobe. In this role, he was a supply assessor and visited makers of armor, swords, saddles, and other equipment to establish suitability for the Royal Armory (Johnson and Green, 1993).

About 300 years later Samuel Pepys, as one of his many appointments associated with the navy, was the Surveyor-General of the Victualling Office where he proved himself an energetic and zealous reformer of abuses intended to sell the Admiralty short, by ensuring him that the goods to be supplied to the ships were of the requisite quality before they were purchased (Johnson and Green, 1993).

It was during World War I that quality took to the air and led the

Royal Aircraft Establishment to try to improve the reliability of British engines. When an enemy engine failed, the prevailing wind usually allowed the pilot to return behind his own lines to fight another day, whereas the Allied pilot forced into the same action finished the war as a prisoner.

After the Armistice, there was a significant change in the scale and diversity of industry in general. Companies evolved from small, self-contained units into integrated operations where individuals no longer had total control over the end product. Individuals were now responsible for a specific part, which would then be passed from operator to operator or firm to firm, gathering other components on the way to completion.

This change was the introduction of inspectors who, independent of the manufacturing operations, would assess the work and return anything that was defective for rectification. Rework and reinspection were here to stay.

For many years this iterative process of make, inspect, accept, or rework has been the basis of the manufacturing industry. It is only recently that the more efficient and cost-effective concept of *getting it right the first time—every time* has started to replace it.

The further expansion in industrial and technological change, which was attendant on World War II, saw an increase in complexity in the manufacturing process and its products.

The first attempt to standardize quality was in the United States of America (USA) where expansion and its effects were greatest and the most significant. This standardization gave rise to MIL-Q-9858, which is a quality system specification, and MIL-I-45208 which specifies inspection system requirements (Mil Spec, 1956). Both standards are still current and are utilized in American defense contracts and elsewhere (Duncan, 1986).

These two standards formed the basis for a series of standards designed for use within the North Atlantic Treaty Organization (NATO). These were called the Allied Quality Assurance Publications (AQAP) 1, 4, and 9. AQAP-1 was a quality system specification and AQAP-4 and -9 were inspection system specifications. The former covered manufacturing, inspection, and testing, and the latter covered final inspection only (Levy, 1993).

Despite its membership in NATO, the United Kingdom did not accept the AQAP. Instead it introduced a series of three similar specifications called Defense Standards (DEF. STAN.).

The most significant difference between the DEF. STAN. and the AQAP was the introduction of some requirements for design to the quality system specification DEF. STAN. 05-21, which otherwise compared with AQAP 1. The other two DEF. STAN. 05-24 and 05-29, were inspection system standards and covered the same subject matter as AQAP 4 and 9, respectively (Breitenberg, 1993).

The Ministry of Defence would assess companies engaged in defense contracts or which were subcontractors to defense contractors, and those found compliant with the requirements of the appropriate DEF. STAN. were registered. In theory, only registered firms could be used for defense contracts. This is an example of second-party assessment, because only two parties, the company and the Ministry, were involved and approval only indicated fitness to meet Ministry of Defence requirements.

At a later date the AQAP were aligned with the DEF. STAN. and progressively Ministry of Defence assessments have been aligned with AQAP standards. The DEF. STAN. are now obsolete.

The AQAP are very militaristic in their content and wording and make considerable use of that misunderstood word *materiel* on which many quality managers' reputations for literacy has foundered. In fact, it is a perfectly proper word, which was introduced in France during the Napoleonic Wars to indicate everything necessary to fight a battle or wage a war except the men and horses. By extension *materiel* now means everything needed to run a business except the personnel.

Within industry at large there was also a need for quality standards to work. Early attempts to meet this need in Britain resulted in standards such as BS 4891 and BS 5179. These were in the nature of codes of practice and had no application in contractual situations.

The solution, which came forth in 1979, was the first edition of BS 5750. This standard was firmly based on AQAP 1, 4 and 9 and was in three parts: 1, 2, and 3. These parts mirrored the AQAPs closely, even to the extent of Part 1 being a quality system specification and Parts 2 and 3 being inspection system specifications.

Like the AQAP, these standards were very subjective and contained a large number of explanatory, nonmandatory notes. Also, like the AQAP,

Parts 1, 2, and 3 were supported by commentaries, called Parts 4, 5, and 6, which contained interpretive material.

This first version of BS 5750 was used not only in a contractual sense between buyer and seller, but as a third-party registration scheme whereby, as an independent organization, it could register companies complying with the requirements of the appropriate part of the standard on behalf of all customers, actual and potential.

The situation that has been described for Britain existed to a greater or lesser extent throughout the world. As a result, a committee of the International Standards Organization (ISO), under the chairmanship of Canada, worked to produce an international quality standard. It considered many national inputs and in 1987 produced a series of standards largely based on BS-5750, its notes and commentaries. This series comprises ISO 9000 which embraces ISO 9001, ISO 9002, ISO 9003, and ISO 9004. In their entirety these standards are as follows:

ISO 9000. *Quality Management and Quality Assurance Standards—Guidelines for Selection and Use.*

ISO 9000-2. *Quality Management and Quality Assurance Standards—Part 2: Generic Guidelines for the Application of ISO 9001, ISO 9002, and ISO 9003.*

ISO 9000-3. *Quality Managements and Quality Assurance Standards—Part 3: Guidelines for the Application of ISO 9001 to the Development, Supply, and Maintenance of Software.*

ISO 9001. *Quality Systems—Model for Quality Assurance in Design/Development, Production, Installation, and Servicing.*

ISO 9002. *Quality Systems—Model for Quality Assurance in Production, Installation, and Servicing.*

ISO 9003. *Quality Systems—Model for Quality Assurance in Final Inspection and Test.*

ISO 9004. *Quality Management and Quality System Elements—Guidelines.*

ISO 9004-2. *Quality Management and Quality System Elements—Part 2: Guidelines for Services.*

ISO 10011-1. *Guidelines for Auditing Quality Systems—Part 1: Auditing.*

ISO 10011-2. *Guidelines for Auditing Quality Systems—Part 2: Qualification Criteria for Quality Systems Auditors.*

ISO 10011-3. *Guidelines for Auditing Quality Systems—Part 3: Management of Audit Programs.*

ISO 10012-1. *Quality Assurance Requirements for Measuring Equipment—Part 1: Management of Measuring Equipment.*

Vision 2000 A. *Strategy for International Standards Implementation in the Quality Arena During the 1990s.*

ISO 8402. *Quality Vocabulary.*

Not all these standards are certifiable standards. In fact, most of them are considered to be guidelines and aids for the quality system. The certifiable standards are: ISO 9001, ISO 9002, and ISO 9003. These are the standards that most people think of when discussing the ISO. The certifiable ISO 9000 series of standards has several outstanding features.

- It is obvious that the standards have been produced by people who are acquainted with the problems and failures that occur in industry and the clauses address these points in a manner that is largely objective. There are only a few notes in the standards. These are nonmandatory. There are no supplementary commentaries.
- Although there are three standards, each is a specification for a quality system. This is in contrast to the three parts of BS-5750, which are progressive and additive. No longer is there a quantum leap from Parts 2 and 3 to Part 1. If the requirements for the production process are added to those of Part 3 and some very minor changes are made in the wording, Part 2 results. If the requirements for design/development and servicing are added to these with similar changes in wording, the standard is converted to Part 1.
- There is little that is dictatorial in the standards. Only rarely do they prescribe a condition. Often, they require the company to establish its own procedures.
- To a great extent, the standards have drifted from the traditional confines of the metal-cutting industry and can be applied, with minimal interpretation, to any industry at large. Some diverse

examples are: food processing, automotive, electronics, medical device, and service industries.

- Although ISO 9004 addresses quality-related cost considerations and product safety and liability, there is no reference to these topics in any of the current operating standards. However, plans for their inclusion in future editions are well underway. Examples are in the area of safety, reliability, product liability, cost of quality, and others.

Although there has been widespread acceptance of ISO 9000 throughout the world, the third-party application of it varies widely. Britain, in 1990, had the most extensive third-party registration system in the world, and more companies were registered by third parties than in any other country. At the same time, Japan had no national third-party registration at all. In the United States, according to CEEM Information Services, as of October 1993 there were 1804 registered companies and the number was climbing (CEEM, 1993).

Many other countries have approved, instituted, or set up third-party registration systems with national recognition. It should be noted that one country, Britain, will register companies in other countries that want to trade in Britain and find that registration by a British organization is advantageous. Mutual recognition of one country's registration system by another is being extensively and increasingly negotiated, but progress is slow and, in many cases, complicated.

In 1991, there were 15 independent bodies in the United Kingdom assessing companies worldwide on their ability to meet the requirements of ISO 9001, 9002, or 9003 and registering those that did. A common standard of assessment is assured by a body set up under the auspices of the government's Department of Trade and Industry called the National Accreditation Council for Certification Bodies (NACCB). More bodies are awaiting accreditation as certifying organizations.

The five largest groups with the scope to undertake assessments of companies in a wide range of industries are as follows:

1. British Standards Institution Quality Assurance. This is a commercial company that receives no public funding at all. It is associated only with the partially publicly funded standards

(making and selling) organizations at the highest administrative level.

2. Bureau Veritas Quality International.
3. Lloyds Register Quality Assurance.
4. Det norske Veritas Quality Assurance Ltd.

    These four accredited certifying bodies each has its origin in well-established, internationally operating, third-party inspection organization.

5. Yarsley Quality Assured Firms Limited. This is an assessment organization that derives from the Institute of Physics via the Fulmer and Yarsley laboratories.

In the United States the number of assessment bodies is ever-increasing. The following are some of them: AT&T, Underwriters Laboratory, QSR, SGS, NSF, Handley-Walker.

Other companies exist to assess companies in specific areas of activity, such as steel reinforcing, electrical wiring, and ceramics. Specialization seems to be the way of the future, with each major industry defining some of the specific applications of the standards to its own needs.

These organizations are in commercial competition with one another, and although the NACCB ensures that common standards are applied, the details of the method and fees vary from one to another.

Many companies were registered to the requirements of BS 5750: 1979. They were given some 15 months to revise their systems to those of the 1987 standard. This period has now expired and all companies are now registered to the new version of the standard. In 1994 the first ISO 9000 revision became effective. Although minor changes were the result, the certification bodies and registrars will certify companies from now on to the new standard. The old certification will be upgraded as the surveillance audits are followed through.

Not only have the standards in the ISO 9000–9004 series been accepted in the national systems of many countries throughout the world, they have been adopted as Euronormes by the European Community and numbered EN 29000–29004. In consequence of this, BS 5750: 1987 is triple-numbered as a British, European, and international standard. Thus, BS 5750: Part 1: 1987 is also EN 29001-1987 and ISO 9001-1987.

In the British system ISO 9000 and 9004 (EN 29000 and 29004) have been combined into a single standard, BS 5750: Part 0 Sections 0.1 and 0.2, respectively.

In 1990, BSI introduced BS 5750: Part 4: 1990 entitled *Guide to the Use of BS-5750: Part 1, Part 2, and Part 3*. This standard is exclusive to the United Kingdom, and it is essential that it is not confused with ISO 9004. It has no relationship to any ISO standard.

It has already been stated that BS 5750: 1979 was very subjective and a lot of guidance was needed for its use. In addition to the textual notes and the commentaries, further guidance was given by BSI's Quality Assessment Schedules, BVQI's Quality Standard Supplements, and Lloyds' Quality System Supplements.

Each was oriented toward a particular industry and contained additional requirements relevant to that industry that were not present in the standard or gave information on how the standard was to be understood in that industry.

In the United States these standards have been termed Q 9000, Q 9001, Q 9002, Q 9003, and Q 9004. They are the same in content as the British, European, and international standards; however, they differ in language.

In addition, the revised 1994 standards provide 21 notes that, even though they are not certifiable, do provide some guidance in helping one to understand and/or implement the standards.

These documents are always under review to make sure that all requirements for quality are covered and are identified in an unbiased language. Furthermore, they are reviewed with an intent to make the standards applicable and workable to as many industries as possible.

Finally, what are the advantages and disadvantages for a company when it implements a quality management system to the appropriate ISO standard or section of BS 5750?

ADVANTAGES

- The company can become more profitable. Poor quality costs British industry some 10 billion pounds each year. Typically, the cost of not finding out what the customer needs and then meeting those requirements faultlessly, every time, is costing companies

15% to 50% of sales. The average profit made by a British company is 4% of sales and any reduction in quality costs goes straight to profits. Companies operating good-quality management systems can reduce quality costs below 5% of sales (Allcock and Unsworth, 1991).

In the United States these costs have been estimated to be anywhere from 10% to 50% of gross sales. Improvements have been made in cost of quality in various industries and published results have shown that some individual companies operate with quality costs as low as 3% of their gross sales (Crosby, 1985).

- Staff is a motivated staff for a number of reasons:

  1. They are not being continually harassed to rework work and still meet the deadline nor are they on the receiving end of the customer's wrath.
  2. The standard requires procedures to be established. When the workforce is involved in this process, they are committed to the standard.
  3. Through its independent audit and management review clauses, the standard opens beneficial horizontal and vertical lines of communication.
  4. The standard requires corrective actions to be identified, taken, and verified as effective. Since the ongoing surveillance activity ensures that these requirements are complied with, frustration is eliminated.
  5. The increased profitability is conducive to job security.

- When approved and registered, the details of the company appear in the Department of Trade and Industry's *Register of Approved Firms* and are also circulated by the certifying body. These publications have a wide and general circulation and can bring additional business to registered firms by bringing their abilities to the attention of potential customers.

- Registered companies are permitted to use an approved firm's logo under a wide range of circumstances, including their advertising. It may not be used in any manner to deceive or imply that a product is approved. This can also attract new business.

  Registration to ISO standard or BS 5750 applies only to

the quality system that the company operates. The third-party assurance that a product conforms to its appropriate standard or manufacturing or performance regulations is provided by the BSI kite mark or safety mark or some similar indication such as the Health and Safety Executive's BASEEFA mark, British Telecom's green disc, and various other marks issued by a variety of organizations.

The operation of a quality system registered to ISO 9000 or BS 5750 is a prerequisite of some product marks, for example BSI's kite mark.

- Although not yet proven by case law, there is a strong consensus of legal opinion that a disciplined quality management system, according to the relevant ISO 9000 standard and the maintenance of records that the standard requires, will provide the best possible plea of mitigation in any product liability case. Many insurance companies providing product liability indemnity policies offer reduced premiums to registered organizations (Kokla and Scott, 1992).

- Registration to ISO 9000 or BS 5750 is a necessary prerequisite to supply to a growing number of government, national, and private organizations.

- The ISO equivalence of the standard gives international recognition of ability.

DISADVANTAGES

- The implementation of a formal quality system is very demanding of resources. Although it is improbable that any company that has traded successfully in a competitive market for any length of time is without a system, the formalization and documentation of the system is time-consuming and may involve considerable secretarial expense. Assessment and registration are costly.

- Unless carefully planned, the system can become non-cost-effective and burdensome.

- By exposing people's actions and cherished practices as not conducive to quality, the process can become traumatic.

- The need to change attitudes and accept new working practices may strain the management capability of the company beyond its ability to manage.

* Although registration of a company to ISO 9000 or BS 5750 is often expected to eliminate second-party assessment, this does not occur in practice. It is possible that more purchasers will accept the standard as the sole qualification for a supplier as it becomes better known and more widespread. However, it is more likely that it will be used as a guide to indicate those companies that are worth looking at more closely.

INTRODUCTION OF THE ISO 9000 STANDARDS

The standards were developed by the International Organization for Standardization, a Geneva, Switzerland, organization founded in 1946 to promote the development of international standards and related activities, including conformity assessment (testing, inspection, laboratory accreditation, certification, quality system assessment, and other activities intended to assure the conformity of products to a set of standards and or technical specifications), to facilitate the exchange of goods and services worldwide. That is why the organization chose *ISO* as the prefix for its numerical standards. It comes from the Greek root *ISO* which means *equal*. It is not an acronym for the International Organization for Standardization.

ISO is composed of nearly 200 technical committees and its members are from over 90 countries, with the U.S. member being the American National Standards Institute (ANSI). The jurisdiction of the standards extends to all areas except those related to electrical and electronics engineering, which are covered by the International Electrotechnical Commission (IEC). The results of ISO's technical work are published as International Standards or Guides. Technical Committee 176 (ISO/TC176) on Quality Management and Quality Assurance began in 1979 on the ISO 9000 standards and it was first approved in 1987.

The detailed description of the development of the standards has been given elsewhere (Peach, 1992; Lamprecht, 1992; Cottman, 1993; Mac-Lean, 1993; BPP, 1992; Voehl et al., 1994). Therefore, in this section we are giving a very limited overview in a pictorial format. Figure 1.1

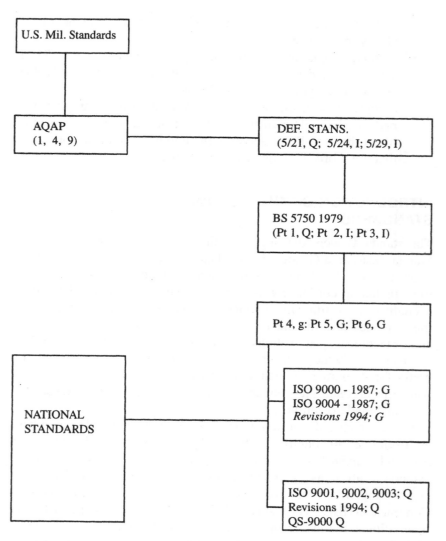

Figure 1.1. Evolution of the ISO 9000 Series standards. Q, quality system; I, inspection system; G, guideline.

shows the evolution of the ISO standards, Figure 1.2 shows the development of the standards, and Figure 1.3 shows the ISO 9000 certification hierarchy.

As we have seen, the ISO 9000 series is the result of an evolutionary process and is composed of primarily five documents (ISO 9000–9004). The series describes three quality standards, defines quality concepts, and gives guidelines for using international standards on quality systems. Contrary to what many believe, ISO 9000 does not apply to specific products and does not guarantee that a manufacturer produces a *quality* product (Stamatis, 1992). The standards are generic and enable a company to assure (by means of internal and external third-party audits) that it has a quality system in place that meets one of the three published standards for a quality system.

An issue usually discussed in relation to the ISO is the question of: "Why the ISO standard?" The answer is a multiple one. However, fundamentally it has to do with the creation of the single Western European Market. After long international negotiations, 12 separate markets became one. Most barriers to free movement of goods (free trade) are being removed. This fact alone presents an unprecedented level of opportunities, as well as considerable changes, for the organizations that want to participate. To effectively trade with one another, and be assured that goods and services meet a consistent set of standards, agreed-upon quality standards had to be developed or accepted within the community. The impetus for international standards was created by this movement toward a free market.

The European market is made up of 18 countries: 12 from the European Union (EU) (called the European Community until 1994), representing 320 million consumers, and six from the European Free Trade Association (EFTA), representing 30 million consumers. Table 1.1 identifies the member countries. By contrast, the United States represents 250 and Japan 120 million consumers, respectively (Hagigh, 1992). These countries have developed bridging agreements or have agreed to 76 technical directives to enable the formation of this free trade market. The European Committee for Standardization (CEN) was mandated to develop harmonized versions of the ISO 9000 series, which are designated EN 9000. All 16 member nations of CEN must adopt EN 29000 as national standards.

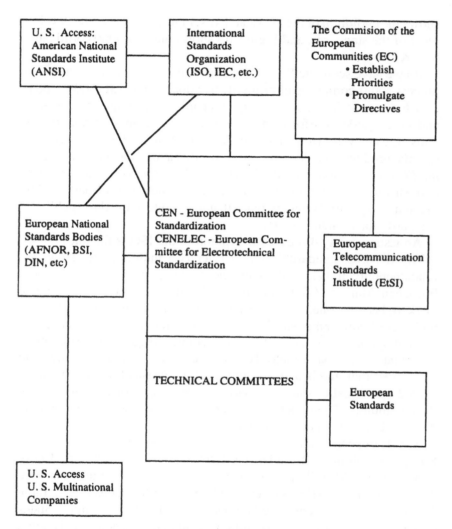

Figure 1.2.   Development of European standards.

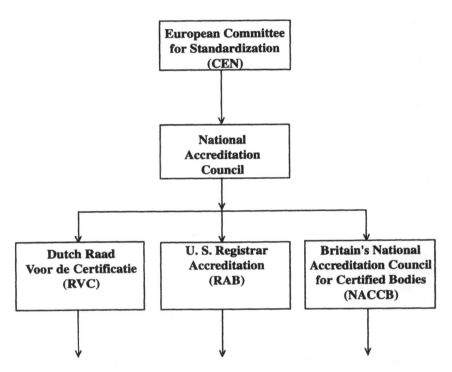

Figure 1.3. The ISO 9000 certification hierarchy. (CEN members are national standardization organizations.)

The standardization has had a profound effect in the United States since the European Community has always been one of our premier customers (Linville, 1992). In 1990 we exported over $98.1 billion in goods and services; therefore, we have a tremendous interest in the world market. To date 91 countries have signed an agreement to abide by these standards.

In the United States the movement has gained ground in several industries, such as steel, chemical, automotive, and electronics, and the U.S. government. The goal of this fast-paced movement is to allow individual organizations and governmental agencies the opportunity to:

Define the quality system that is appropriate and applicable to a given organization.

Table 1.1   The European Market (350 Million Consumers)

European Union (EU)	European Free Trade Association (EFTA)
Belgium	Austria[a]
Denmark	Finland
France	Iceland
Greece	Norway
Ireland	Sweden
Italy	Switzerland
Luxembourg	
Netherlands	
Portugal	
Spain	
United Kingdom	
Germany	

[a]Austria was accepted as a EU member as of January 1, 1995.

Demonstrate, to the customers, the commitment and management system to maintain quality.

Compete in the international markets.

Follow standard safety and product liability regulations and or procedures.

Reduce cost and provide a practical, results-oriented target(s).

Help the organization maintain quality improvement gains.

Minimize supplier surveillance—through second-party audits.

Provide a platform from which to launch a "continuous improvement" program such as total quality management (TQM), Malcolm Baldrige awards, and so forth.

Involve *all* employees by stimulating understanding of quality systems.

REFERENCES

Allcock, T., and Unsworth, T. (June 1991). *Lead Auditor Course: Training Manual*. BSI, Detroit, MI.

Breitenberg, M. (April 1993). *NISTIR 4721: Questions and answers on quality, the ISO 9000 standard series, quality system registration, and related issues*. U.S. Department of Commerce. National Institute of Standards and Technology, Washington, DC.

Bureau of Business Practice (1992). *ISO 9000: Handbook of Quality Standards and Compliance.*

CEEM Information System (October 1993). *Quality Systems Update.*

Corns, M. C. (1968). *The Practical Operation and Management of a Bank*, 2nd ed. Bankers Publishing Co., Boston.

Cottman, R. J. (1993). *A Guidebook to ISO 9000 and ANSI/ASQC Q90.* Quality Press, Milwaukee, WI.

Crosby, P. (1985). *Quality Improvement Through Defect Prevention.* Philip Crosby Associates, Winter Park, FL.

Duncan, A. J. (1986). *Quality Control and Industrial Statistics*, 5th ed. Irwin, Homewood, IL.

Durant, W. (1954). *Our Oriental Heritage.* Simon & Schuster, New York.

Goetz, A. (1973). "The laws of Eshunnana." In James B. Pritchard, Ed. *The Ancient Near East, Vol. 1.* Princeton University Press, Princeton, NJ.

Guerdan, R. (1956). *Byzantium: Its Triumphs and Tragedy.* George Allen, New York.

Hagigh, S. (Feb. 24, 1992). "Obtaining EC product approvals after 1992: what American manufacturers need to know." *Business America.*

Johnson, R., and Green, D. (1993). *Lead Assessor Training.* P. E. Batalas, Singapore.

Kokla, J. W., and Scott G. G. (1992). *Product Liability and Product Safety Directives.* CEEM Information Services, Fairfax, VA.

Lamprecht, J. L. (1992). *ISO 9000: Preparing for Registration.* Quality Press, Milwaukee, WI.

Levy, M. P. (1993). *20 Questions and Answers on the ISO 9000 Standards.* Document No. Q101, rev. 1. Quality Systems Resource Facility, Troy, NY.

Linville, D. (Feb. 24, 1992). "Exporting to the European Community." *Business America.*

MacLean, G. E. (1993). *Documenting Quality for ISO 9000 and Other Industry Standards.* Quality Press, Milwaukee, WI.

Meek, T. J. (1973). "The code of Hammurabi." In James B. Pritchard, Ed. *The Ancient Near East*, Vol. 1. Princeton University Press, Princeton, NJ.

Military Specification. (1956). *MIL-Q-9858: Quality Control System Requirements.* Superintendent of Documents, Washington, DC.

Peach, R. W., Ed. (1992). *The ISO 9000 Handbook.* CEEM Information Services, Fairfax, VA.

Stamatis, D. H. (August 1992). "ISO 9000 standards: are they for real?" *ESD Technology.*

Voehl, F., Jackson, P., and Ashton, D. (1994). *ISO 9000: An Implementation Guide for Small to Mid-sized Business.* St. Lucie Press, Delray Beach, FL.

2
Quality Vocabulary

In any field of endeavor there is a special language that facilitates understanding through specificity and meaning. The ISO world is no different. The following definitions are based on the ISO 8402. The purpose of this chapter is to facilitate some of the most common jargon in the ISO series and to offer some further commentary on the interpretation of specific words and phrases. Note that only a selected number of terms have been identified; for the exact and complete vocabulary see the actual standard, which is given in ANSI/ASQC A8402-1194.

DEFINITIONS

Concession Waiver.   Written authorization to use or release a quantity of material, components, or stores already produced but which do not conform to the specified requirements.

Defect.   The nonfulfillment of intended usage requirements.

Design Review. A formal, documented, comprehensive, systematic examination of a design to evaluate the design requirements and the capability of the design to meet these requirements and to identify problems and propose solutions.

Grade. An indicator of category or rank related to features or characteristics that cover different sets of needs for products or services intended for the same functional use.

Inspection. Activities such as measuring, examining, testing, and gauging one or more characteristics of a product or service and comparing these with specified requirements to determine conformity.

Nonconformity. The nonfulfillment of specified requirements.

Product Liability; Service Liability. A generic term used to describe the onus on a producer or others to make restriction for loss related to personal injury, property damage, or other harm caused by a product or service.

Production Permit; Deviation Permit. Written authorization, prior to production or before provision of a service, to depart from specified requirements for a specified quantity or for a specified time.

Quality. The totality of features and characteristics of a product or service that bear on its ability to satisfy stated or implied needs.

Quality Assurance. All those planned and systematic actions necessary to provide adequate confidence that a product or service will satisfy given requirements for quality.

Quality Audit. A systematic and independent examination to determine whether quality activities and related results comply with planned arrangements and whether these arrangements are implemented effectively and are suitable to achieve objectives.

Quality Control. The operational techniques and activities that are used to fulfill requirements for quality.

Quality Loop; Quality Spiral. Conceptual model of interacting activities that influence the quality of a product or service in the various stages ranging from the identification of needs to the assessment of whether these needs have been satisfied.

Quality Management. That aspect of the overall management function that determines and implements the quality policy.

Quality Plan. A document setting out the specific quality practices, resources, and sequence of activities relevant to a particular product, service, contract, or project.

Quality Policy. The overall quality intentions and direction of an organization as regards quality, as formally expressed by top management.

Quality Surveillance. The continuous monitoring and verification of the status of procedures, methods, conditions, processes, products, and services, and analysis of records in relation to stated references to ensure that specified requirements for quality are being met.

Quality System. The organizational structure, responsibilities, procedures, processes, and resources for implementing quality management.

Quality System Review. A formal evaluation by top management of the status and adequacy of the quality system in relation to quality policy and new objectives resulting from changing circumstances.

Reliability. The ability of an item to perform a required function under stated conditions for a stated period of time.

Specification. The document that prescribes the requirements with which the product or service has to conform.

Traceability. The ability to trace the history, application, or location of an item or activity, or similar items or activities, by means of recorded identification.

COMMENTS

Quality

These comments on the preceding definitions are offered for further explanation and/or clarification of the specific terms. They are offered in the spirit of making the ISO standards more user-friendly and more easily understood. They are not by any means to be used instead of the actual word and/or definition as defined in the ISO standards.

1.   The term *quality* is not used to express a degree of excellence in a comparative sense nor is it used in a quantitative sense for technical evaluations. When it is necessary to identify the quality in these terms, then the specific and appropriate adjective should be used. For example:

Relative Quality.   Should be used where products or services are ranked on a relative basis in the *degree of excellence* or some *comparative sense*.

Quality Level, Quality Measure.   Should be used where precise technical evaluations are carried out in a quantitative sense.

2.   In a contractual environment, needs are identified and/or specified, whereas in other environments, implied needs should be identified and defined where appropriate.

3.   Where possible, the needs (specifications/requirements), wants, and expectations of the customer ought to be identified and defined.

4.   Quite often we must recognize that our needs can change over time. This implies periodic revision of requirements. Generally, the flow of change is as follows:

Needs become obsolete.
Wants become needs.
Expectations become wants.
Other expectations become our expectations.
The cycle repeats.

5.   Needs are usually translated into features and characteristics with specified criteria based on "the" specific customer. These needs may include—but are not limited to—aspects of usability, safety, availability, economics, environment, reliability, and maintainability.

6.   Product and/or service quality is always influenced by many stages of interactive activities (negotiation) and optimization of related activities, such as: design, production, service operation, maintenance, financial, and so forth.

7.   We must always be aware that the economic achievement of satisfactory quality involves ALL stages of the quality loop as a whole. It does not matter that sometimes we do identify some stages but not others. When we do identify the stages, we do so only for emphasis. For

example: quality *attributable* to design or quality *attributable* to service or quality *attributable* to implementation or quality *attributable* to production, and so forth.

8. Remember that quality overall is defined by the customer. Unless the needs, wants, and expectations of the customer are satisfied and/or are exceeded, there is no quality. However, some other definitions may also be appropriate, such as: fitness for use (the Juran traditional definition); the price of conformance/nonconformance (the Crosby definition); continual improvement (the Deming definition), and so forth. All these definitions represent "some" facet of quality, and in all cases a fuller explanation is required. While differences must be recognized, the emphasis of quality is—or ought to be—on the functional use/cost relationship.

Grade

1. Grade always reflects a planned and/or recognized difference in requirements. The emphasis is on use/cost relationship.

2. Where grade is identified numerically, it is usually common for the highest grade to be 1 and the lower grades to be 2, 3, and so forth. If the grade is identified by a point score or a pictogram, then the lowest grade usually has the fewest points or pictures, respectively.

3. A high-grade product or service may be of inadequate quality as far as satisfying needs, and vice versa.

Quality Policy

The quality policy is only one element of the corporate policy and is authorized ONLY by top management.

Quality Management

1. Quality management includes strategic planning, allocation of resources, and other systematic activities for quality, such as quality planning, operations, and evaluations.

2. The responsibility for quality management belongs ALWAYS to top management. This is contrasted with the attainment of desired quality, which requires the commitment and participation of all members of the organization.

Quality Assurance

1. Quality assurance within an organization acts/serves as a management tool. On the other hand, in contractual situations, quality assurance also serves to provide confidence to the customer.

2. Quality assurance requires continual evaluation of factors that affect the adequacy of the design, specification, verification, and audits of production, installation, and inspection operations. To provide confidence implies that producing evidence for effectiveness may be a requirement.

3. Quality assurance is complete ONLY when the requirements reflect the needs, wants, and expectations of the user.

4. Quality assurance always focuses on and emphasizes the planning of quality as opposed to the appraisal of quality.

Quality Control

1. Quality control involves operational techniques and activities aimed both at monitoring a process and at eliminating *root* causes of unsatisfactory performance at relevant stages of the quality loop in order to result in economic effectiveness.

2. When one is referring to quality control, special care should be taken in describing what really is meant, so that confusion is avoided. For example: Is manufacturing quality control the same as company-wide quality control?

3. Quality control focuses on and emphasizes the appraisal of quality as opposed to the planning of quality.

Quality System

1. One of the most important issues in any quality system is the complexity, length, comprehensiveness, and so forth. The quality system should only be as complex, comprehensive, and so forth as needed (appropriate and adequate) to meet the quality objectives.

2. Remember that the objectives are defined by the organization. So, in the quality system, you tell the world what you are going to do.

3. For contractual, mandatory, and assessment purposes, demonstration of the implementation of your quality system and its elements may be required. (For ISO certification, the demonstration is a must and is conducted by a third-party assessor.)

Quality Audit

1. While the quality system tells the world what the organization is going to do, the audit makes sure that what you say is actually what you do. Quality system audits are performed for ISO certification.

2. To keep the integrity of the system whole, the audit must be conducted by staff who do not have direct responsibility in the areas being audited.

3. Quality audits can be conducted for internal or external purposes.

4. Audits have many purposes including, but not limited to: management information, corrective action, benchmarking your organization for future evaluation on improvement, and so forth.

5. Under no circumstances should a quality audit be confused with or used in lieu of *surveillance* or *inspection* activities performed for the sole purpose of process control or product acceptance.

Quality Surveillance

1. Quality surveillance may be carried out by or on behalf of the customer to ensure that the contractual requirements are being met.

2. Surveillance may have to take into account factors that can result in deterioration or degradation with time

3. For ISO certification the surveillance is mandatory by the registrar and is performed either once or twice a year depending on the registrar. The registrar also has the right to come in to the organization at any time with or without preannouncement.

Design Review

1. *All* design reviews by themselves *do not* ensure proper design.

2. A design review may be conducted at any stage of the design process.

3. The participants at each design review MUST be qualified and appropriate for all pertinent functions affecting the design at hand and the quality of the product.

4. The focus of the design review will depend on the objective and scope of the product. Generally, some of the most common items reviewed to assess capability of the design are as follows:

Fitness for purpose
Feasibility
Manufacturability
Measurability
Performance
Reliability
Maintainability
Safety
Environmental aspects
Life-cycle costs

Traceability

1.   The term *traceability* has at least three main meanings.

- In a distribution sense, it relates to a product or service.
- In a calibration sense, it relates measuring equipment to some standard (primary-physical constant(s); properties; national; international).
- In a data collection sense, it relates calculations and data generated throughout the quality loop to a product or service.

2.   Traceability requirements should be as specific as possible for some stated period of time and point of origin.

Change Waiver or Letter of Deviations

1.   Make sure that the waivers are limited quantities, or periods for specified uses.
2.   ALL waivers should be signed by the appropriate authorized person.

Reliability

1.   Sometimes the term *reliability* is used to denote a probability of success or a success ratio.
2.   The field of reliability is always reviewed, and therefore, the definition may change (see IEC Publication 271; ISO 8402 for updates).

Product Reliability

1.   It is of paramount importance to be aware that the limits on liability do vary from country to country according to their national legislation.

2.   The European Union (EU) is in the process of developing a series of product liability directives. As the directives are developed and implemented, the world of product liability will be somewhat standardized.

Nonconformity

1.   *Nonconformity* refers to the absence and or departure of one or more quality characteristics or quality system elements from specified requirements.

2.   The basic difference between nonconformity and defect is that specified requirements may differ from the requirements for the intended use. (Compare item 1 under "Defect").

Defect

1.   *Defect* refers to the absence and/or departure of one or more quality characteristics from intended usage requirements.

2.   See item 2 under "nonconformity."

Specification

A specification should refer to or include drawings and ALL relevant documents and should also indicate the means and the criteria whereby conformity can be verified.

REFERENCE

ISO 8402:1994. *Quality Management and Quality Assurance—Vocabulary.*

3

Quality System Requirements

There are many tools and techniques to improve controls of the process and a whole series of approaches to teamwork and improvement issues to develop people skills. ISO 9000 helps to provide the infrastructure for the systems side of TQM and, in more general terms, a system of quality.

This chapter addresses the intent of the standard clause by clause and provides a meaningful interpretation of each of the clauses.

The ISO 8402 defines quality assurance as: *All those planned and systematic actions necessary to provide adequate confidence that a structure, system, or component will perform satisfactorily in service.* The essence of this definition then requires us, the *purchaser* (the company, the user), to have a set of rules that gives us *confidence* when we purchase a product and or service from a *supplier*, that it will consistently meet our needs. These set of rules are applied to a supplier to give assurance that their systems work.

This system is driven to the subcontractor base with the expectation

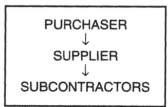

Figure 3.1 Confidence in suppliers.

and confidence that quality will be controlled and optimized at all levels. This push is illustrated in Figure 3.1.

One of the most serious questions, then, is the issue of how to manage quality. Note that the concern is on management, not control. Figure 3.1 demonstrates that your quality confidence is only as good as the incoming quality from your suppliers and their subcontractors. Furthermore, for that quality to exist, the organization must define it. The organization's management is in control.

It is a common fallacy that the quality systems relate to production alone. All areas of an organization can use and benefit from the use of quality systems, just as all organizations (whether manufacturing or service) can use and benefit from these systems when a product or service is produced with the intent for a *fit for purpose* and supplied to a customer when the customer wants it and at an economical price. This economical price is not necessarily the *cheapest* or the most *expensive*, rather it is the *best value* price as viewed by the customer.

In deciding on the method of application of ISO 9000 to a company, it is important for the organization (1) to consider the specific requirements placed on the organization related to regulated industries and (2) to understand the specific jargon of the standards.

1. A regulated industry is an industry that produces products for which the EU Commission has developed or is developing an EU-wide technical harmonization directive. It provides manufacturers with a single set of requirements for products offered for sale in the EU. Currently, some of the directives (a partial list) are:

Toys
Simple pressure devices

Personal protection equipment
Construction products
Machines
Gas appliances
Product conformity testing equipment
Medical devices
Telecommunication terminal equipment

The implication for the specific industries with directives is that they have to follow specific and, in some cases, additional requirements. For example, the food, software, automotive, nuclear, and other industries may use the ISO 9000 as a framework for their quality system; however, additional specific requirements may be necessary to comply with the regulated industries.

Guidelines on these specific requirements are in development and they are published and updated on a continual basis. The registrar of your choice can provide you with the additional information, if required.

2.   The usage of jargon, on the other hand, will facilitate the implementation and the understanding of the standards. Of special interest are the following words:

May: When the standard addresses something with *may*, it implies the things that would give quality system developer benefits if incorporated in the documentation.

Should: When the standard addresses something with *should*, it implies the things that are specific to an industry and are essential to an effective quality system.

Shall: When the standard addresses something with *shall*, it implies the things you are obliged to define and carry out. It may be directed to a specific standard or document.

KEY FEATURES OF THE ISO 9000

The ISO 9000 series is a standard produced as five documents, i.e., ISO 9000, ISO 9001, ISO 9002, ISO 9003, ISO 9004. Specifically:

The ISO 9000 is a set of Quality Guidelines for selection and use of quality management and quality assurance standards.

The ISO 9001 Quality Systems is a model for quality assurance in design/development, production, installation and servicing.

The ISO 9002 Quality Systems is a model for quality assurance in production and installation.

The ISO 9003 Quality Systems is a model for quality assurance in final inspection and test.

The ISO 9004 Quality Management and quality system elements is a complete set of quality guidelines.

Only the system requirement standard ISO 9001, ISO 9002, and ISO 9003 are written in mandatory language that allows the documents to be used in contract conditions. Therefore, these are the only standards that a given organization may consider for certification. On the other hand, ISO 9000 and ISO 9004 are not written in a mandatory language; rather they are written as guidelines to select the appropriate system and to define the elements of what the system should cover, respectively.

The use of numbers does not indicate differing degrees of excellence. That is, ISO 9001 certification is not better than ISO 9002, which in turn is not better than ISO 9003. They are system requirements for specific applications. When a company undertakes innovative design and when its customers are likely to require assurance of design capability, then it inevitably must operate to ISO 9001. However, if a company produces products against established specifications or provides a service against a written requirement for service delivery, then ISO 9002 may be adequate to meet those needs.

The emphasis of the international standard is on the management of quality systems. All the clauses of the standard are relevant and important to the company's effective operation, but certain key features highlight the strength of the document in terms of correcting the cause of problems and getting all inputs correct with an efficiently functioning process to ensure that outputs are meeting expectations.

THE REQUIREMENTS

When one considers the requirements of the ISO 9001, it is useful to think of three distinct areas: management responsibilities, company-wide activities, and specific requirements.

In this section the focus is only on the management responsibilities.

For the company-wide activities and an overview of specific requirements, see Chapter 5 and Chapter 6.

Section 1:   Scope

In this section the standard identifies the intent of the standard: "The requirements specified are aimed primarily at achieving customer satisfaction by preventing nonconformity at all stages from design through to servicing."

This section is not a certifiable requirement.

Section 2:   Normative Reference

In this section the standard provides the sources and references used in the standard.

This section is not a certifiable requirement.

Section 3:   Definitions

In this section the standard identifies the sources of the definitions used in the standard as the ISO 8402. In addition, it identifies three additional ones:

Product: Result of activities or processes. This product can be tangible or intangible or a combination of both. Of importance in the delineation of the product definition is the explanation as it differs from the American National Standard (ANS). For the ANS the term *product* applies to the intended product offering only and not to unintended *by-products* affecting the environment. This differs from the definition given in ISO 8402.

Tender: Offer made by a supplier in response to an invitation to satisfy a contract award to provide a product.

Contract; accepted order: Agreed requirements between a supplier and customer transmitted by any means.

This section is not a certifiable requirement.

Section 4:   Management Responsibilities

This is the actual certifiable portion of the standard. The following is a detailed explanation and interpretation on a per element basis.

Clause 4.1:   Management Responsibility.   This is a very extensive clause, made up of three subsections, each of which highlights particular activities that senior management must undertake. It is significant that this is the first clause in the standard and it demands that the senior management organize and plan for quality. In other words, the quality responsibility cannot be delegated to the lower levels of management. Senior management, is always responsible and must be dedicated to quality. Their commitment to quality must be beyond reproach and always exemplary rather than merely involved.

The significance of the clause addressing the senior management as a first requirement establishes right up front the need for sincerity and commitment. The standard recognizes that without this sincerity and commitment the quality system of the organization will not work. It takes management to allocate the appropriate resources, and without the sincerity and dedication, other priorities will be emphasized at the expense of quality.

As part of this dedication and commitment they must always reinforce with words and deeds what quality is all about and what it really means to the organization. It is their responsibility to preach improvement and consistency.

Clause 4.1.1:   Quality Policy.   Before anyone in any company can hope to be consistent in any activity, there must be a policy. In this policy objectives must be enumerated, so that the policy becomes an instrument of direction and vision for the company.

It is not enough to say, "This is our policy or these are our objectives." They must be documented. Documentation implies collaboration. This collaboration may be any *two* sources. For example: two individuals, reports, procedures, and/or in formal writing. It is of paramount importance to recognize that the ISO 9000 standard does not require anything in writing. However, it requires documentation; therefore, the written form of documentation becomes the most expedient and accurate for the system.

Since this commitment to documentation is important and necessary, management must find a way to tell everyone in the organization what the policy and objectives are and how they affect each of them individually and collectively. This means finding a way to *cascade* the understanding down the organization and at *each level* making sure that management at that level has installed the system as well as providing for maintenance of the same.

Specifically, clause 4.1.1 requires a company to state its objectives as well as its policy for quality.

Objectives are the quantified goals that are specified and then attained. Knowing where the destination (goal) is allows managers to plan the journey to quality.

While the policy is formulated by the most senior group of managers, company objectives for quality will require the participation of many managers as their efforts influence these objectives. There are some benefits in this:

- The act of defining company objectives helps to unify managers.
- Company objectives are necessary prerequisites to operating on a planned basis rather than from crisis to crisis and department to department.
- Defining company objectives enables a subsequent comparison of performance against objectives.

Objectives in quality may be a reduction in scrap, complaints, rework, returns, and so forth. Or they may merely define a certain performance.

To make the system improve, the objectives preferred should be those that will improve some aspect of the company's performance, delivery, lead-time, response, reliability, and so forth.

Clause 4.1.2: Organization

Clause 4.1.2.1: Responsibility and Authority. This requirement is concerned with making sure that the various tasks in an organization are clearly defined and people's authority to manage, do, or check work is also clearly defined. This definition of authority should be traceable to senior management.

After the quality policy and objectives have been defined and agreed on in writing, the way is cleared for the next stage of the process. This is often accomplished by preparing schematic organization charts showing technical and functional interfaces and may include job descriptions for the positions shown on the chart, if the company wishes to use that method. It is strongly recommended that in the organizational chart positions should be listed, as opposed to names. The reason for this is that people move more often than the organization changes. As a consequence, fewer changes will be required.

The detail is based on necessity. To outsiders it has to be clear that the organization is defined, and to insiders it must be clear exactly what their job is and the limits of their authority and responsibility. Furthermore, the definition of authority must demonstrate the relationship of the quality authority as it relates to senior management.

Clause 4.1.2.2: Resources. An important feature of any control activity is the objectivity of the verification and the independence of the person verifying. The bias of any impropriety must be eliminated. Doubt of objectivity must be removed. The standard requires that management examine what work is being done and ensure that, where they think it necessary, there is a means of verifying—it has been done the way they wanted it done.

Clause 4.1.2.3: Management Representative. The senior management shall appoint someone as a representative to ensure that the standard is being followed and to assist in its implementation and effectiveness. It does not matter what other duties this person may have as long as he or she is in a position with the defined authority and responsibility. There must be a direct *link* or *traceability* to the senior management.

Within the duties outlined it is necessary to define the management representative. Quite often, this position is filled by the Quality Manager, but anyone nominated by the senior management and given the authority and responsibility can provide this function. When that happens, this management representative is called the ISO 9000 facilitator or the ISO 9000 coordinator. The facilitator may or may not be a full-time position. The purpose, the duties, and/or the functions of this facilitator may cover several and diverse activities. Some typical activities are:

Internal audits
Inspection

Statistical methods
Quality/inspection planning
Corrective action
Product liability
Calibration
Supplier certification or surveillance
Defect prevention programs
Customer complaints analyses
Monitoring the improvement
Quality costs

The list is not meant to be exhaustive. Neither is it mandatory. It merely gives some examples of traditional duties associated with the function of the ISO facilitator (the management representative).

Clause 4.1.3: Management Review.   Once the system is installed, management is responsible to review its effectiveness and appropriateness. It must be remembered that the *system* belongs to management since management has the responsibility to define it and generate the policy of operation as well as the objectives, which of course all together establish the system. It is of paramount importance that this review take place at specific intervals, because without it the quality system will stagnate.

To avoid this stagnation the review must take either a formal or informal format as long as critical information is reviewed and the process is documented. Management reviews generally include, but not limited to:

Assessment of the internal audit results
Review of corrective actions
Review of suggested changes
Monitoring of the system

Clause 4.2:   Quality System

Clause 4.2.1:   General.   This clause requires that the management formally defines the activities necessary for a quality system in more detailed documents. Typical documents are the quality manual, procedures, instructions, and or standard operating procedures (SOP). The reason for the detail is to put this documentation into practice and to make sure it is working effectively.

Of special interest is the term *specified requirements*. It is used for the first time in the standard and it means any requirements an organization is required to work to:

- Contract requirements specified by a customer in a contract and agreed to by the supplier
- The policy, objectives, procedures, and instructions that the management define, which will be used to do the work in the company
- The quality management system standard, i.e., ISO 9001, ISO 9002 or ISO 9003
- Regulatory requirements
- Directives in product or service lines

Clause 4.2.2:   Quality Procedures.   This clause specifies that the procedures of the organization should be consistent with the requirements of the ANS.

Clause 4.2.3:   Quality Planning.   This clause, which defines how the requirements of quality will be met with appropriate documentation, is specific in identifying eight items that qualify as quality planning; however, the note of the same clause identifies the appropriateness of such documentation. Therefore, it is implied that the eight items given in this section are only recommendations and each organization should develop its own. The recommendations of this clause are:

- The preparation of quality plans
- The identification and acquisition of any controls
- Ensuring the compatibility through the design, production, installation, servicing, inspection, and test procedures
- Suitable verification
- Clarification of standards
- Keeping quality records

Clause 4.3: Contract Review

Clause 4.3.1: General. In order for a supplier to meet customers' requirements, the supplier must be sure what those requirements are and know that they can be met when the order is taken. This means reviewing each order and providing a record of the review.

Clause 4.3.2: Review. It is essential that if a contract is to be undertaken between two parties, both parties are absolutely clear and in agreement on what is to be supplied. Only by reviewing all the contract requirements fully can a company be sure that it will satisfy its customer.

The key steps are as follows:

- Ensuring that the requirements are clearly understood and agreed upon by both parties and that these requirements are recorded and kept on file.
- Reviewing all the requirements internally to ensure that the resources, organization, and facilities are available and capable of meeting all technical and commercial requirements.
- Reviewing any subsequent changes to determine the effect of meeting the requirements.

Although the standard is written around the situation that involves a tender (purchase agreement) document typical of a large project, the principles are just as true when purchase is made of a simple catalog item. Someone has to ensure that the customer's requirements of item, delivery, and so forth can be met.

It is not always recognized that the content of sales literature that is sent out to many customers constitutes a contract when purchase is made on the basis of those claims. ISO 9001 can have the desirable advantage of causing such literature to state what can be done as against what would be liked to be done.

Many of today's companies utilize a checklist in the order entry system that requires certain items of information to be supplied before the order can be processed. Although this is relatively easy on a computer system, manual methods for defining the specification in this way can also be very effective.

Clause 4.3.3:   Amendment to Contract.   This clause demands appropriate documentation, when and if amendment(s), modifications, and changes are issued to the contract.

Clause 4.3.4:   Records.   All records that are used as part of the contract review should be maintained.

Clause 4.4:   Design Control.   To think that a single clause in a standard would adequately cover all the requirements, is presumptuous. This clause identifies the minimum requirements of design control that a company has to define formally and how it complies with these requirements.

The complexity of this clause is demonstrated by the nine subdivisions, some of which are further divided to define specific requirements of design.

Clause 4.4.1:   General.   The management of the design function has to decide how the design work is going to be done and, furthermore, how the work is going to be checked and the design verified.

The general requirement is for a method of controlling the design activity and verifying that the design meets the requirements. The requirements may be customer-specific, industrial-specific, and/or regulation-specific.

The rest of the subclauses specifically address the details of the requirements of a sound design.

Clause 4.4.2:   Design and Development Planning.   Planning is essential to quality in any task. This demands that a plan, perhaps in the form of quality function deployment (QFD), a block diagram, reliability calculations, failure mode and effect analysis (FMEA), and design reviews, has to be prepared that references the responsibility for doing and the responsibility for verifying at each stage. As the design evolves, the plan will need to change to reflect the changes in the design. The plan therefore becomes "dynamic" and always ready to accommodate the changed design.

Furthermore, this plan must cover the concerns of how the company will keep up to date in both design and development. The plan must be descriptive. It has to communicate to someone who is not familiar with the design. At least one way of handling this situation is to have a procedure and instructions covering the specifics of the design.

Planning of the design must take place and specify the responsibilities for each part of the design. The parts may be in the form of: (a) system, (b) subassembly, and (c) component.

Clause 4.4.3: Organizational and Technical Interfaces. The essence of this clause is to assure that the inputs for the design are cross-functional and multidisciplined in nature. The standard recognizes that any designer does not work alone. This implies that for the design to be at optimum other departments—i.e., production, sales, purchasing, servicing, inspection, and others—must have an input to that design. Information, therefore, needs to be kept under control, relevant, and current so that the status at any point in time can be established and only valid design decisions are made.

The standard is concerned with the means of ensuring that information necessary to different groups is properly and adequately received and reviewed. Therefore, the mechanism for controlling that information is important and needs formal definition by management.

Clause 4.4.4: Design Input. All inputs necessary to a design have to be identified, reviewed by authorized persons for validity, and documented. The issue here is *documentation*.

The review of these documents must establish that incomplete and/or ambiguous requirements are resolved between all input sources and the designer(s).

To prove that a system exists, appropriate documentation must exist. Minutes of meetings, memos, design reviews, and other pertinent information do qualify for appropriate documentation; however, they must be kept and updated as needed.

Clause 4.4.5: Design Output. The design function produces something that can be made and tested and something that complies with what the customer requested.

Design output is in the form of drawings, calculations, specifications, analyses, and so forth. These should refer to the design inputs that determined the requirements and also state or refer to the acceptance criteria; i.e., what defines that this product will be correct?

Within the specification, ALL requirements must be addressed—legal, regulatory, and customer requirements and any part of the design that is crucial to the safety and proper functioning of the product.

The design has to be finished at some point and be deemed complete. That basis must be identified and stated in the form of the media required, of product features, of calculations and analyses, and so forth. That basis must relate directly to input and must conform with the appropriate regulatory requirements (stating how) and identify characteristics that relate to safe and proper functioning of the product.

Clause 4.4.6: Design Review. The people involved in the design work have to be qualified. That does not mean that everyone has to have a *journeyman's card* or a *college degree*. It means that management has to decide what qualifications and/or experiences and/or training is appropriate and adequate for the designated task. All design personnel should have the appropriate credentials and/or certifications for their respective tasks. These should be properly and appropriately documented with either the design department and/or the human resources department.

The definition for the credentials and/or certifications may be found in a job specification, job description, and/or personnel records. In some cases some of these requirements may be found in procedures and/or department guidelines. The actual location of these sources will depend on the organization.

In essence then, this clause emphasizes the need for identification as to who does what, who is responsible for verifying each part of the design, what are the qualifications of personnel, how to define the qualifications, how the qualifications were formalized, and how to keep them current.

Clause 4.4.7: Design Verification. A means of verifying the design must be defined. The list provided in note number 10 is not, nor is it meant to be, exhaustive. Rather the list provides examples of what a company may consider as design verifications. Some of these examples are: design reviews, alternative calculations, and comparisons with similar proven designs.

Checking or verifying of the design must take place against the requirements agreed upon at the start and as the design evolves.

Some companies do this by a series of design reviews within the project team with perhaps attendance at various times by external people. This is done at various stages or milestones during the design project and is often done against the characteristics of reliability, maintainability, inspectability, and durability. Other companies may utilize computer aided design (CAD) systems, simulations, prototypes, or technical audits by independent persons of the design who are technically knowledgeable to examine and evaluate the design.

As part of the verification process, others may use alternative methods of calculation (reliability engineering), testing, FMEA, or any other available tool that is appropriate for the particular design.

Many designs are developments and/or modifications of previous designs—designs that in the past have demonstrated a capability of performance. When these designs utilize common principles, the data can be used as acceptable for verification.

All of this requires examination. The complete process of verification has to be a formal one.

Clause 4.4.8: Design Validation. This clause demands validation to ensure that the product conforms to all of the customer's requirements. Notes 11–14 provide some clues as to how the validation is to be performed.

Clause 4.4.9: Design Change. The changes made to the design must be clear, must be documented, and must show a level of review consistent with the original preparation of the various parts of the design, i.e., the same authority.

Any design is likely to change before it is complete. Before change is permitted, the following types of questions need to be addressed:

- Does the product still conform to the original requirements?
- Is its fitness for use affected?
- Can it still be made, tested, etc.?
- Will it affect safety or operation?
- Is there anything else that would be affected?

Clause 4.5: Document Control

Clause 4.5.1: General. This clause establishes the need for the company to have and maintain documented procedures to control all documents and data that relate to the requirements.

Clause 4.5.2: Document Approval and Issue. Documentation and paperwork generally are not in short supply in most companies. What is usually not available is the correct documentation (appropriate, adequate, up-to-date, with the obsolete documentation properly removed). Correct documentation implies that documentation in use has been approved by the right people.

One of the most common areas of deficiencies in quality assurance systems is the control of documents. The focus of document control is to ensure that the right version or issue of a document is available.

This means that for the main documents in particular, there exists within the organization a list that shows who has what information and its location. This information can enable the appropriate person in charge of document control to send the changes or a new copy of the changed document to the appropriate person in a reasonable time.

The standard implies that only authorized documents should be in use. Therefore, a system of approval should be stated. Of special interest is the temporary changes. All organizations have the need for temporary changes, perhaps allowing certain authorized persons to make handwritten changes to a document. This is acceptable as long as it is in a defined and authorized procedure and the persons affected are aware of it.

Any document that undergoes a great deal of change will ultimately need to be renewed completely, i.e., redrawn. There can be no set rule for this, but if there is a danger of confusion because of multiple changes, the document needs to be reissued.

Clause 4.5.3: Document and Data Changes. Documentation that needs to be changed must be changed provided the change goes through the same authority that approved the original document. When a document is changed, persons using the document must be able to find out what the change is.

Clause 4.6:   Purchasing

Clause 4.6.1:   General.   The requirement for this clause is to identify that the company placing a purchase from a supplier or subsupplier is responsible for ensuring that whatever is bought is correct. This implies that the company must have in place systems that assure the conformance of the product and/or service purchased based on the requirements set forth in the contract to purchase.

Clause 4.6.2:   Evaluation of Subcontractors.   The standard requires that the company actively considers the reasons for choice of subcontractor and can show evidence of the reason for the choice. The decision must be based on what the buying company considers important and is able to show that these aspects are covered.

The requirements here demand that a sound basis is established for the use of subcontractors and that the information given to them is clear.

The basis for selection may be by various means. It may be that as a policy the company wishes to use only suppliers who have been assessed and certified by an accredited third party. This policy may, of course, limit the choice of suppliers. However, the company may require it as a condition of doing business and may insist that within a given time all their suppliers must gain certification. This, of course, has been at least one of the reasons for the growth of registration.

Another basis for use of a supplier is an assessment by the company itself, by a regulatory agency, by a specific contractual agreement of the customer, by a product assessment, or by any combination of these. The criteria for the selection should be established by the company and it should take into account the degree of assurance necessary from that supplier.

If there is some history of satisfactory supply from a supplier, this can provide the evidence necessary for acceptance. Such an assessment may be the result of incoming product inspection, delivery performance, after-sales service received, response time to complaints, corrective action, follow-up, and so forth.

In the absence of registered suppliers through third-party registration, it is required and strongly encouraged that the company have a formal certification program that identifies the suppliers in a tier system, such as:

Certified—the most advanced self-certification. Most of the purchases are done with this kind of supplier. The relationship is looked upon as a win-win.

Preferred—the status is less than the certified. The supplier has the potential of becoming a certified supplier if it follows the quality systems prescribed by the customer.

Approved—the suppliers without historical data. These are usually the ones with whom minimum interaction takes place.

Depending on the certification level attained by the supplier, the company may accept the product and/or service from the supplier based on a monitoring program. The monitoring will make sure that the criteria established as being important are followed and meet the requirements. The monitoring may be implemented through statistical process control (SPC), MIL-STD, incoming inspection, and other statistical formats.

Clause 4.6.3: Purchasing Data. This is one of the most explicit requirements in the standard. It requires absolutely full and clear detail, which must appear on the purchasing document.

The information sent to the supplier as an order should be considered as external work instructions. Orders and supplements should be under approved document control and should describe exactly what is necessary in order for the supplier to comply.

The requirements for the content of a purchase order can be specified as a procedure. In that way it can be learned by those preparing orders and can also be checked.

Clause 4.6.4: Verification of Purchased Product

Clause 4.6.4.1: Supplier Verification at Subcontractor's Premises.
This requirement addresses the concerns of companies that are accustomed to visits by customer representatives to the suppliers or the company. It must be clear that when a customer inspects the product, this in no way absolves the company from its responsibility for purchasing the *correct* product.

Clause 4.6.4.2: Customer Verification of Subcontracted Product.
During purchase from a supplier the company's customer may wish

to verify at the supplier's works or the company's works that the products purchased comply. If this is an important condition to the company, then a clause in the purchasing agreement may be necessary to formalize this right.

Regardless of who makes the purchase and how it is verified, neither the company nor the customer nor the supplier is absolved from its obligations under the contract between them.

Clause 4.7: Control of Customer-Supplied Product. When a customer supplies material or product to be incorporated into the final product by the supplier, the supplier must ensure that the material is correct, is stored and handled correctly, and so forth. In those situations where a customer supplies material to be incorporated into the final product, a company must have a way of dealing with this. The best way is to include the checking, handling, storage, etc., of this material under the control system for the company's own supplied product.

Clause 4.8: Product Identification and Traceability. This requirement is particularly flexible to encourage the principle of adding unique identities to a product to tie it to specification or drawings, etc. The degree of traceability necessary and even possible varies extensively across different industries.

In highly safety and/or critical industries, traceability back to the original source is maintained by unique identity numbers on the product itself. Lesser requirements may be for traceability back to where it was purchased only and in the case of certain products, textiles for example, the country of origin.

The necessity for recall of a product will determine the extent of traceability. When safety is not an issue, the costs involved will be from either the customer's or the supplier's point of view.

During manufacturing, identification is often attained by accompanying documents by batch, lot, or any other appropriate means of identification.

Clause 4.9: Process Control. This requirement deals with manufacturing and installation and is very broad in its content.

The control requirement may include aspects of process approval by qualification and monitoring. Whatever method is used, it must be clear what constitutes success or otherwise. This clause explains the manufacturing function and contains one of the most contentious statements in the standard.

The requirement states the need to formally plan the stages of production necessary; furthermore, it requires the processes to be "carried out under controlled conditions" including: "documented procedures defining the manner of production, installation and servicing, where the absence of such procedures would adversely affect quality. . ."

The requirement quite reasonably is that written instructions are necessary everywhere—except where they are not needed. Only management can decide when a procedure must be formal.

It must be made clear to all personnel in the organization what it is that they are required to do. The written word is perhaps the most common, efficient, objective way of transmitting this information.

Some of the reasons for making a procedure formal are the following:

1. If it is a specified requirement, i.e., part of a contract, a requirement by the manager of the department, or a requirement of the quality systems standard, then it must be formal.
2. If there is a difference between what the manager who is responsible thinks should happen and the staff who have to operate the procedure think, then a written procedure will help to overcome that difference. However, it does not ensure it. A difference in understanding between persons may be overcome by written procedure—but it is not a guarantee.
3. Where inputs to a procedure are made by different people, perhaps in different departments, then a written procedure may help in ensuring a common understanding.
4. When it is necessary historically to understand the basis for records, a written procedure can help to provide the information.
5. If is necessary to bring in people who do not usually operate that procedure, a written procedure can obviously help.
6. When staff decide themselves that the only way they can be sure they know what to do is if they write it down—even unofficially—then there is an established need for a written procedure. In this case there is probably a high likelihood that they will write the procedure.

The implication of the preceding statements is that management must demand that appropriate education and training for the people involved be available and must encourage them to participate. Once the training has been established and implemented, then management must provide the system to monitor and evaluate that those involved are doing what they were trained to do.

Further requirements within this clause allow the company to consider the use of various process controls or checking of product features to maintain control and in all cases to provide to personnel a clear specification of success, a written standard, a picture, a representative sample, and so on.

Certain processes, by their nature, can only by controlled by predefined requirements being followed by personnel qualified in some particular skill. Inspection of the finished product would not be effective in determining whether there were deficiencies.

Such processes require a degree of prequalification and strict monitoring against prescribed procedures. Examples are welding, forging, casting, soldering, plastic molding, certain food and drug manufacturing, software coding, etc.

Clause 4.10: Inspection and Testing

Clause 4.10.1: General. All inspection and testing activities should be documented and properly maintained.

Clause 4.10.2: Receiving Inspection and Testing. Items purchased from suppliers must not be used unless they are correct (4.10.2.1). They must meet the contractual requirements. A degree of goods-received inspection may therefore be necessary if this is the only way that confidence is gained about the supplier's ability to supply to order.

It may be totally unnecessary to carry out any inspection other than the identification and counting, if the supplier is certified through either a third-party registrar or the company's certification program. On the other hand, it may be essential to very thoroughly (100%) inspect everything. Only the management can decide (4.10.2.2).

It is acceptable to release material from receipt into production if there is a means to recall it, should it subsequently be found to be incorrect(4.10.2.3). This, however, could be an expensive risk to take. It should be noted that proper identification and traceability (clause 4.8) will help in the recalling process, if indeed it is needed.

Clause 4.10.3: In-process Inspection and Testing. During the manufacturing process the management must ensure that the product is checked, either by direct measurement or by monitoring of the process. It must not allow the product to proceed unless the required inspections have been carried out and the product has been approved.

During manufacturing there are stages at which it is prudent to check the product, whether by first-off approval or by inspection before very costly processing is done.

These stages need to be defined in the manufacturing documents in the control system and need to prevent material from proceeding to these operations unless shown to be acceptable.

Clause 4.10.4: Final Inspection. Final inspection is defined as the ultimate check before the product is shipped. The requirement, therefore, for final inspection is to determine that all previous inspections and tests have been carried out, records are available, and the product has been shown to be acceptable.

In the final inspection, which is performed prior to release for delivery to the customer, a company must be sure that all the tests necessary have been done satisfactorily and that the available information shows this. At each stage of inspection, there may be product that does not conform. This needs identifying as such.

Clause 4.10.5: Inspection and Test Records. In this clause the requirement is to provide formal evidence of the inspection and/or tests against the criteria that formed the product specification.

Clause 4.11: Control of Inspection, Measuring and Test Equipment

Clause 4.11.1: General. This is one of the most detailed clauses in the entire standard. Essentially the equipment used for measuring acceptability of the product has to have appropriate procedures and documentation of control, calibration, inspection, measurement, and testing, to demonstrate conformance and capability of product to the specified requirements.

Clause 4.11.2: Control Procedure. If inspection or test of a product is carried out to determine its acceptability to specified requirements, then the inspection or testing equipment must be capable of giving a valid measurement. The following controls are necessary:

1. Registry of equipment. All equipment used to measure and test needs to be listed. However, they may not all need to be calibrated. The list needs to determine which must be and which need not be calibrated.

2. Plan. Clearly a rolling plan is necessary so that all equipment does not require recalibrating at the same time. Not all equipment will need the same frequency of calibration. Any portable equipment in constant use is likely to need more frequent calibration than pieces used in a controlled environment.

3. Accuracy tolerances. A piece of equipment used to make a measurement is chosen based on the product tolerance. The measuring equipment itself has errors. When these errors could give an invalid product measurement, then the measuring equipment is not capable. These tolerances and errors need to be determined.

4. Instructions and records. Calibration has formal needs for procedures and records of the checks. Actual values attained are required. "OK" or a *checkmark* is not acceptable as a record.

5. Traceability to *national* standards. The standards used to check measuring equipment must be related to the ultimate standard. When this is not possible, then agreement with the customer is usual.

6. Indication of calibration status. It must be possible to determine the status of each piece of equipment. This can be accomplished by affixing labels or using color codes, but this is not always possible.

7. Ensure environmental conditions. Make sure that all environmental conditions are suitable for the appropriate testing to be carried out.

8. Ensure accuracy. Make sure all conditions are appropriate for accuracy and fitness for use.

9. The standard requires that within the manual or procedures a company has a method for deciding what it will do if a piece of

equipment is found to be out of calibration and unknowingly it has probably *passed* product that is unacceptable.

Clause 4.12: Inspection and Test Status. This is one of the most self-explanatory clauses of the standard. Where product is likely to have different status—i.e., not inspected, inspected and passed, inspected and rejected, held, etc.—then the status must be clear. Tags, labels, or accompanying documentation can provide the means of status identification. Particular locations may also facilitate the identification process, as well as specific containers specified for unusable items or scrap.

The context of this requirement is related to products, but the principle should be inherent throughout the system. The requirement is to make it clear that an item is unchecked, checked and accepted, or checked and rejected so that only properly approved items will be used.

Clause 4.13: Control of Nonconforming Products

Clause 4.13.1: General. As a result of inspection *at any stage* product may be deemed as nonconforming. It has to be prevented from being used until a decision has been made about it. It must be marked and it must be documented. Whenever anything occurs that does not conform with specified requirements, people need to be made aware of it.

Clause 4.13.2: Review and Disposition. After something wrong is found, someone needs to decide what to do about it and with it. Who can decide and what can be decided must be defined in writing.

Only certain authorized people can decide what to do with the defective product (defined in the quality manual and or procedures). If it can be used *as is*, then a waiver is raised. If the customer agrees it can be used as it is (when it affects a customer requirement), then a waiver must be raised with the customer. If it is to be reworked to bring it back into specifications, then a different procedure has to be raised from the standard procedure. If it can be downgraded and used only in certain limited applications, then it must be identified and some instruction raised so that it can be controlled. If none of these situations exists, the product must be scrapped. To prevent it from being inadvertently used, perhaps it needs to be damaged in a major physical way.

Clause 4.14: Corrective Action

Clause 4.14.1: General. Anything found that is wrong or potentially wrong must be corrected. In order to do that, it is necessary to determine the root cause. The essence of this clause is to establish documented procedures for both corrective and preventive action. If a company is wondering whether to put certain systems in and not bother about others, then this requirement should be considered as one to have above all others.

Clause 4.14.2: Corrective Action. Any product nonconformance, customer return, complaint, system deficiency, etc., is a failure of the company's management system. Any company wishing to correct any of the problem(s) identified needs to record them, determine the root cause(s)—rather than the symptom—identify alternative solutions, select the appropriate solution, and monitor the solution for improvement and consistency.

Good records of product and process characteristics help to identify potential problem areas, and this principle is part of the requirement in the standard. Therefore, all nonconformances from whatever source need to be collated and analyzed.

Clause 4.14.3: Preventive Action. This clause provides some guidance as to what constitutes preventive action. It is of paramount importance to recognize that the standard is not prescriptive in what kind of corrective and/or preventive action the company will define. Rather, the standard defines as a requirement the need and demonstration of a system that is appropriate, applicable, and useful in the definition and solution of a problem.

Clause 4.15: Handling, Storage, Packaging, Preservation, and Delivery

Clause 4.15.1: General. At all stages throughout a company, product is moved, stored, packed, preserved, and then delivered, and each of these stages has its own dangers. This clause establishes the need for appropriate documentation to accommodate all stages identified. Specifically:

Clause 4.15.2: Handling. Different products require different handling. Cleanliness is crucial in the food industry but impossible in iron founding. Delicate electronic items require electrostatic precautions and the use of special materials in contact with products, etc. Whatever the requirements, these must be formally defined in general or specific procedures and made to work.

Clause 4.15.3: Storage. Storage must maintain items in the condition in which they entered the storage area. Security may be important in some companies with only certain people authorized to enter stores. Of course, storage applies to any parts or items held between stages of processing, so total security may be impossible. Control over passage in and out of storage is essential by use of authorized symptoms and personnel. The principles of "first in, first out" stock checking and stock taking are features of good storage practice. Separation of visually similar items with specified nonconformances is a practical necessity.

Clause 4.15.4: Packaging. Packaging must protect the product until it reaches the customer. If used to sell the product and present it, it needs to last longer. Usually there are standard packaging requirements for any product unless a customer specifies differently.

Clause 4.15.5: Preservation. Preservation methods must be identified and appropriate documentation must be in place to assure that the product is appropriately segregated and protected.

Clause 4.15.6: Delivery. While the extent of responsibility may vary from contract to contract regarding delivery, the company needs to ensure that the delivery method is consistent with protection needed; e.g., certain items can be carried in the open air, others must be protected from damp, dirt, shock, etc.

Clause 4.16: Control of Quality Records. Quality records provide evidence that the product meets requirements and that the quality system has operated effectively. In most cases, each procedure and work instruction needs to contain, reference to the documents used and the record that is provided to show effective operation of the system. Although the standard calls them quality records, they will in most cases be the ordinary documents and records produced in each department of the company.

Clause 4.17: Internal Quality Audits. In order to provide verification evidence of the systems operation, audits need to be carried out of all the activities outlined in the quality manual. The audits are carried out against checklists, which are samples taken from the requirements of instructions, procedures, quality manual, or standard. The actual operation and effectiveness of those requirements are then checked in practice.

Any area that is not being operated as the procedure specifies is recorded as deficient. At the end of the audit, a report is written and is distributed to the department heads of the participating departments and the management. It is hoped that a corrective action will be issued to change the way the process works and/or the written procedure. The responsibility of an internal audit is usually delegated to the quality assurance department. However, it is not essential that audits are performed by quality assurance. Everyone in the organization is encouraged to participate.

An internal audit is a verification activity, and therefore, it must be assured that independence exists between the auditor and the auditee. This essential factor can be accounted for by having a representative of another department do the audit. It is good practice to allow many people to carry out internal auditing as it increases people's knowledge about the company as a whole and removes the interdepartmental barriers that are so often erected in companies. The audit can be a powerful management tool and is undertaken in much the same way that an audit by a second party or third party would be done.

Audits are required to look at the whole system and verify whether there is a system, whether that system is put into practice, and whether it is effective. The audits must react to the needs of the system and include follow-up of any area found to be deficient in any way. Any deficiencies found must be reported to management responsible for the area.

Clause 4.18: Training. The intent of clause 4.18 in general is to establish the requirements for training as a strategic goal. It also establishes the requirements for certifying and recertifying employees involved in performing critical and specialized functions at a given organization.

Specifically, this clause focuses on the following questions:

Does the training program cover quality awareness for the entire organization?

- How are the training needs identified?
- Is the top management included in this training?
- Are the objectives of the training appropriate for what is really required?

Does the training program cover the development of special skills required for various processes, including the qualifications or certification of the personnel where applicable?

- How are the qualifications determined?

   Education?
   Training?
   Experience?
   Combination?

- How is management training and development identified? How is it carried out?

How does the training program cover the requirements for requalification and or (re)certification?

- Is formal education necessary?
- Is a vocational school appropriate?
- Is the apprenticeship program certified?
- Are the requirements for qualification, requalification, and (re)certification appropriate for the skills of the organization?

How are the instructors/trainers selected?

- How do they get qualified?
- Is there a certification program?

What records are kept of the results of the training?

- Are training procedures available and/or controlled? Why or why not?
- Are the results of training monitored and/or documented to show verifiable achievement?
- Are all training records/procedures legible, identifiable, and retrievable?

Are retention times established and recorded?
- How was the retention time established? Is it appropriate? Why? Why not?

Are the procedures, documents, and records available to the customer (if they are contractual agreements)?
Is there a budget allocation for training?

- Is technical training provided (to update skills)?
- Is orientation part of the general training?
- What constitutes *normal* training?

Although clause 4.18 is very specific and focuses on the issues of training of a special nature, the intent and the scope of the paragraph are more than the explicit identification. It covers ALL employees performing routine, critical, and specialized functions related to deliverable items. This includes both the management and nonmanagement personnel. In addition to this general impact, the clause may also include requirements specified by the customer's contact.

As part of the clause 4.18 requirements, it is also essential to look at the responsibility for developing training and testing criteria. The criteria for such an evaluation more often than not can be found in the jurisdiction of the quality management and administration department, in their respective quality manual section.

Typical information relating to appropriate training records are:

Employee's name and identification number
Department of employee
Date of training and duration of training
Checkmark for certification or recertification
Location of training
Kind of training

Name of course
Name of instructor
Certification number
Date of issue of certification
Name of provider of the training
Function of the certification
Expiration date (if applicable)

Typical training concerns may be:

How does this individual participate in or become selected for the required or specified training?
What are the makeup consideration(s)?
How does the verification occur?
How is the scheduling set?
Who sets the scheduling?
Who keeps the regular and/or updated records?

The requirements of clause 4.18 may be frightening to some companies; however, in reality they are nothing more than a substantiation of the quality system that the company itself has defined in its own quality manual. The ingredients identified in the quality manual ought to be: identification of training needs, the training program, and records.

The purpose of clause 4.18 is to tell me what you are going to do, tell me what you are doing, and then, during the physical audit, show me what you are doing. This, of course, is not unique to clause 4.18 but applies to the entire ISO 9000 standards.

Clause 4.19: Servicing. This is a unique clause in the standard because it is conditional upon whether a contract requires servicing. It demands that if it does, then means exist to ensure that the servicing is done as it is meant to be and that there is verification that this is the case. Appropriate proof of existence is a record.

Service personnel need to be kept advised of new products and competitors' information. Service staff are at the *sharp end*, often dealing with frustrated users of the product. Their reports are of real-life situations and in most cases are extremely valuable.

Clause 4.20: Statistical Techniques

Clause 4.20.1: Identification of Need. This requirement is perhaps the most misunderstood clause of the standard. It is assumed by most to mean that specific statistical *charting* and traditional statistical process control (SPC) must be performed. Nothing could be further from the truth.

This clause is intended to encourage management to consider ways of determining process capability and to review current methods of inspection. Furthermore, it is intended to identify appropriate need and use of statistical techniques, i.e., control charts, sampling techniques, descriptive statistics, and anything that is applicable, appropriate, and useful for the organization and, more specifically, for the process.

Clause 4.20.2: Procedures. Any pure quality assurance program can be viewed as the removal of undesirable variables from the process, whatever that process is. In today's environment this principle is essential and can be accomplished in at least two ways: (1) statistical techniques and (2) sampling. Even though both approaches are useful, the traditional sampling tables are not adequate to control products down to parts per million or parts per billion levels. Therefore, other approaches are necessary, and this clause is demanding the establishment of appropriate and applicable methods.

This clause is written as an encouragement to companies to utilize any statistical techniques that will identify, measure, control, and monitor the variation of the process.

Summary

After reviewing the standard, one can appreciate that the ISO 9001 requirements are extremely flexible and extremely extensive. They allow a company to do whatever it wants, provided the company ensures that their way is documented and meets the customer's requirement(s).

The ISO 9001 requires that a company plans for quality and that it does what it says and then records it. It also requires that it audits and reviews what it has done against the objectives and takes action on the difference.

It requires a company to make decisions on quality and it forces honesty and integrity in those decisions. This can only be an improvement over all other systems currently in use among different organizations.

CHANGES IN THE ISO 9000 STANDARDS

One of the mandatory requirements in the structure of the ISO is that the standards must be reviewed at 5-year intervals. The first such revision took place in 1992 and the second is scheduled for 1997.

The 1992 changes were published in 1994 and provide adjustments

Table 3.1 The Hierarchy and Comparison of the ISO Standards

Standard	Clause	Description
ISO 9001 (20 elements)	4.4	Design and R&D
ISO 9002 (19 elements)	4.19	Servicing
	4.9	Process control
	4.6	Purchasing
	4.20	Statistical techniques
	4.18	Training
	4.17	Internal quality audits
	4.16	Quality records
	4.15	Handling, storage, packaging, preservation, and delivery
ISO 9003 (16 elements)	4.14	Corrective action
	4.13	Control of nonconforming product
	4.12	Inspection and test status
	4.11	Inspection, measuring, and test equipment
	4.10	Inspection and testing
	4.8	Product identification and traceability
	4.7	Control of customer-supplied product
	4.5	Document control
	4.3	Contract review
	4.2	Quality system
	4.1	Management responsibility

and clarifications, rather than major changes in the standards. (In this book we use the 1994 revision as our guide in discussing the standards.) The 1997 changes are expected to address issues concerning liability and quality costs, among others.

COMPARISON OF THE CERTIFIABLE STANDARDS

While the ISO 9001, 9002, and 9003 standards provide specific requirements for certification in different situations, all of them have common clauses. Specifically, ISO 9001 has 20 clauses, ISO 9002 has 19, and ISO 9003 has 16.

The 16 clauses in ISO 9003 are found in all the standards and constitute the basis for the structure of the quality system. ISO 9002 includes three more than the ISO 9003, which are included in the ISO 9001. ISO 9001 includes one more clause than ISO 9002 and in total is the most complete and stringent standard. Each clause and its description are listed in Table 3.1.

All three standards begin the certifiable requirements at clause 4.0. The first three sections deal with scope, references, and definitions.

REFERENCE

ANSI/ASQC Q9001:1994. *Quality Systems—Model for Quality Assurance in Design, Development, Production, Installation, and Servicing.* ASQC, Milwaukee, WI.

4

Third-Party Assessments

In this chapter we will explain the place of third-party assessment in the context of quality management systems and give some examples of the methods used by the various certification bodies. Some (limited) consideration will also be given to the international scene with regard to third-party assessments.

OVERVIEW

A third-party assessment is one that is undertaken by an independent body to establish the extent to which an organization meets the requirements of an applicable standard or set of regulations.

Third-party assessment bodies can be used to assess against any required standard. However, in this chapter we will concentrate on assessments of quality systems to the ISO 9000 series.

The independent auditing body—the certification body or registrar—issues a certificate of registration, indicating acceptance of the organization as a *company of assessed capability* or something similar. This certification is issued after a review of the quality system and a physical audit of the organization. A typical certification process is shown in Figure 4.1.

The original certification takes approximately 10–24 months (from preparation to submitting the application for certification). After the certification is achieved, the certification body will visit the assessed organization once or twice a year for surveillance purposes and every 3 years for a major recertification assessment. The cycle of the surveillance and recertification is dependent on the policy of the registrar.

Once this certification is issued, it bears witness to the world at large that the assessed organization complies with all the requirements of the applicable standard. Certification does not guarantee product or service quality. It guarantees only the system.

Certification bodies are normally paid by the company that they assess and they have no other association, including consulting services, with the assessed body. This is why they can demonstrate that the assessment has been carried out without any bias.

Third-party assessment has been a common feature of commerce in the United Kingdom (UK) for many centuries. For example, hall markings began in A.D. 1140. As systems became available it was inevitable that third-party certification would be accepted not only in the United Kingdom but all over the world. Currently, there are about 16,000 companies in the UK and about 3200 companies in the United States, registered to the ISO 9001 or ISO 9002, and the number is increasing at an exponential pace.

In the United States third-party registration is also not totally foreign. The seal of approval from the accounting firms in all annual reports is a form of third-party registration. The college entrance exams (SAT, ACT) and professional/graduate entrance exams (GRE, GMAT, LSAT, etc.) are also a form of third-party registration.

It is not surprising, however, that many of the countries that have accepted the ISO series standards look to the UK for advice and assistance on the implementation of the standards. They do indeed have the most experience in the area, and as a consequence they have become the coaches to the world. This can be seen in countries where no system is yet established, where individual companies turn to UK certification bodies for assessment.

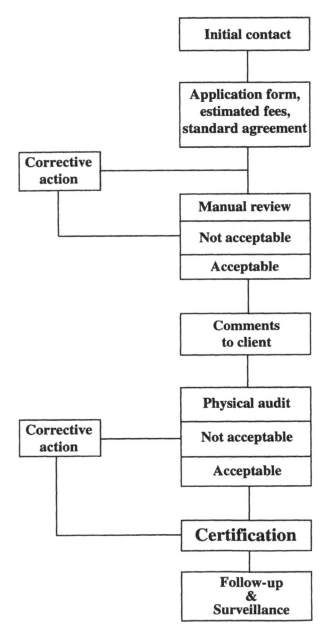

Figure 4.1 A generic ISO 9000 certification process.

THE SYSTEM IN THE UNITED KINGDOM

The National Accreditation Council for Certification Bodies (NACCB), funded by the Department of Trade and Industry (DTI), is the organization responsible for the accreditation and supervision of certification bodies. A certification body that has been accredited by the NACCB is allowed to indicate this by the use of the *Crown and Tick* logo.

Many certification bodies have been accredited for different specialties and industries. In fact, some bodies have accreditation in a wide range of activities, and these are the bodies generally used to assess quality systems to the ISO 9000 standards.

When an assessment has been successfully conducted, the certification body will issue a certificate of compliance to the specific standard, attached to which is a definition of the scope of activities that have been assessed. The assessed company may then indicate its certification by the use of two logos: the Crown and Tick and the logo of the certification body.

The logos are not allowed to be applied to the product, but only to such items as stationary, brochures, etc. They can be applied in certain controlled circumstances, to packaging and decals for advertising purposes.

THE SYSTEM IN THE UNITED STATES

Until recently, U.S. companies relied on quality system registration firms in Europe and Canada to register their quality systems. Today, the number of U.S.-based organizations offering consulting services, assessment, and/or quality system registration is growing rapidly.

One of the reasons for this growth is that in 1989, the Registrar Accreditation Board (RAB) was established as an affiliate of the American Society of Quality Control (ASQC) to develop a program to evaluate the quality of services offered by registrars. RAB issued its first approval in March 1991, and several more firms have been approved since then. The RAB and the American National Standard Institute (ANSI) agreed to form a joint U.S. program in December 1991. In February 1992, RAB announced the establishment of an ISO 9000 auditor and lead auditor certification program.

In addition to the certification programs the RAB is seeking mutual recognition from the EU Council. Preliminary negotiations began on October 20, 1992, and are still in progress as of this writing.

In the meantime, the degree of interest and pressure felt by U.S. manufacturers to seek registration currently varies significantly by industry. In many of the *high-tech* or *high-risk* product areas where product reliability is crucial, the market pressure on U.S. manufacturers to seek registration is likely to be considerable.

THE SYSTEM IN THE WORLD AT LARGE

Many countries (91) around the world have now adopted the ISO 9000-1994 series of standards into their own national system. Of special interest is the European Union, which has adopted the standards as Euronormes (EN) 29000–299004-1988.

The tendency appears to be toward a worldwide recognition of the value of complying with the ISO quality system standards and continued progress toward third-party registration as being a reliable demonstration of compliance.

Most countries of the world have adopted the English system with the exception that they have their own equivalent of the NACCB. It is not uncommon to find that most of the governing bodies in individual countries are controlled by a governmental agency.

ASSESSMENT METHODS

Each certification body has its own method of assessment and certification. However, all of them have some common ground and it is this common ground that we are going to focus on.

The events of the certification procedure are:

1. Initial contact from the organization to the registrar. General information is exchanged and appointments are set for the preassessment meeting.

2. Preassessment visit. Here either a questionnaire is filled out or an actual visit takes place to establish the amount of work needed.

3. Quotation from the registrar. A formal quote for the certification and surveillance services is given to the organization.

4.   Acceptance of the quote. The organization signs the quote or a legal contract for the certification and surveillance as well as the price. The agreement is forged at this stage.

5.   Registrar asks for the quality system. The quality system—sometimes called documented system or desk audit—is requested for review. This review assesses the system of the company as compared with the specific ISO standard that the company is seeking certification in. The purpose of this preaudit is to identify any omissions or ambiguities and/or any major nonconformities before the real compliance audit.

6.   When the quality system is accepted, a request for the compliance audit is made.

7.   The compliance audit is scheduled. It is generally undertaken with a team of auditors (assessors). The team is made up of one lead auditor and three to four auditors. The audit usually takes no more than 5 days. A typical audit has the following elements: opening meeting, audit, and closing meeting.

8a.   Certification is issued. If everything goes well and no nonconformities have been found, then certification is issued. The time for receipt of the actual certificate is 4–8 weeks after the assessment. Although the actual certificate is not in the possession of the organization, nevertheless the organization is certified as of the end of the audit.

8b.   Certification is denied. The denial may be in one of two forms: (1) due to minor noncompliances or (2) due to major noncompliances. In either case the organization might be reassessed when the noncompliances are corrected.

Any nonconformity found must be recorded and its effect assessed by the lead auditor. Some registrars have a guidance procedure to assist in the ultimate decision.

A major noncompliance is the absence or the complete breakdown of a required element of the system. A required element is any of the subsections of section 4 of the applicable standard. Sometimes, the term *major* is used interchangeably with *serious noncompliances* or *hold points*.

A *minor noncompliance* is an isolated failure to comply with specified requirements. A single minor noncompliance would not normally be a cause to fail the registration. However, a series of related minor nonconformities, in the judgment of the lead auditor, will more often than not constitute a breakdown of a procedure and/or the system. At this point, all related minor nonconformities are classified as major.

9. Surveillance. This is the ongoing program of making sure that the organization keeps up with the system. For details see the next section.

10. Appeals. If there is a conflict, the registrar usually has procedures that will define the appeals process. The process is generally as follows:

If, during the audit, an auditor does not conduct himself/herself in a professional manner or some other complaint is justified by the auditee, the auditee has the right of appeal to the registrar. The actual process will depend on the specific registrar.

At that point, the registrar will investigate the complaint ALWAYS with a person independent of the complaint. If the registrar is unable to satisfy the customer, the auditee has the right to complain to the certification body. At this stage, the certification body may review the complaint by asking for additional information and/or may reissue an audit. In any case, it has the ultimate decision. No more appeals are allowed after this.

The registrars have the right to decertify an organization that fails to maintain an adequate standard.

SURVEILLANCE

To receive certification from a registrar is not the end point. Rather, it is a beginning commitment to a quality system that needs monitoring to ensure continued compliance with the standard. The actual monitoring varies from registrar to registrar; however, there are some common points.

All registrars have some kind of monitoring system. Some have a system of regular, unannounced audits; others have a reassessment at regular intervals (3 years is a common one) with either one or two supported audits annually.

Since one of the objectives of the certification is to assure confidence in the system, the idea of the surveillance is to make sure that the effectiveness and the assurance of the system will be continued for the assessed organization.

A typical surveillance may cover the following:

- Check for maintenance of the internal audit program and appropriate corrective action.
- Check customer complaints and their follow-up.
- Check for satisfactory completion of all corrective actions agreed upon at the previous audit (internal or third party).
- Sample check on aspects of the quality system, possibly guided with either records, recorded nonconformities, and/or minor nonconformities from the last audit.
- Check appropriate use of the registrar's logo.
- Check for follow-up on ALL nonconformance items.
- Check whether the internal audits are utilized by top management for *continual improvement* purposes.

5

Implementation Strategy for ISO

This chapter addresses the issues concerning why and how to evaluate the organization as well as to prepare the organization for the implementation process of ISO 9000. Specifically, we are going to address the issues of need assessment and the mechanics of the ISO 9000 implementation process at all levels of the organization.

NEEDS ASSESSMENT

The needs of the organization and the employees in a given company are not static; they undergo change(s) over time. Therefore, it is important that management review themselves. The forces for change are many and are constantly with us, occurring as a result of changes in employee and/or corporate expectations, changes in technology, changes in knowledge, and a variety of changing social patterns.

To adequately assess the perceived needs of a system requires the involvement of representative segments of the corporate culture. By seeking needs information from the representative groups, management can be more responsive to the *wishes* and real *needs* of all the employees. The identification of needs involves a discrepancy analysis (gap analysis, need analysis, etc.) that identifies two opposite positions and the difference, if any, between them. A model of a gap analysis may look like the following:

Perceived   GAP   Perceived
needs →   ← wishes

A typical discrepancy analysis might use the following set of questions:

1. Where are we now?
2. How important is this?
3. How well do you feel this is being done?
4. Where would we like to be?
5. What would it take to get us there?

By comparing the answers to each of the above questions, it is possible to ascertain where significant discrepancies exist and where they do not exist, what we have to do, and what effort we must put forth for the implementation.

Obtaining organizational needs from the employees is not an end in itself, but instead represents a valuable source of information regarding the current status of the corporate system. A needs assessment can provide the corporate system with some of the best available information on the immediate characteristic of the corporate training programs and those areas where attention and resources should be directed.

Needs Assessment Model

A simple model for a needs assessment is presented here based on four phases. If the reader is interested in more detailed information, we recommend the book by Kaufman and English (1979), who present a very thorough model and explanation of needs assessment based on the notion of: *what is* and *what should be*. Their model, although very detailed, is beyond the scope of this book. However, its application can be very fruitful in pursuing a needs assessment in any organization.

Phase 1. Developing the Needs Assessment Framework

Here the corporate goals and objectives are set. Determining the goals and objectives is the primary function of this phase.

Phase 2. Determining Training Discrepancies

Essentially this process involves a comparison of *what is* with *what should be* in three areas of the training setting: employment achievement, program and/or training operations, and preference assessment. A second component of the discrepancy phase is to identify the current operational status and the expected operational status. The most important guideline in determining such discrepancies is to ensure an unbiased, objective reporting.

Phase 3. Setting Tentative Goal Statements

Given the information collected and analyzed in phase 2, management must analyze and interpret these data within the goal framework. Some of the alternatives are:

Validate existing goal statements.
Develop new goals as indicated and supported by the discrepancy analysis.
Refine or revise existing goal statements.
Eliminate outmoded or outdated goals.

Phase 4. Determining the Rank Order of the Goals

The final phase of the process involves determining the priority of goal statements.

Essential to the effective conduct of the needs assessment is the sincere commitment of management at all levels. It is this commitment that will facilitate the process and will bring it to its closure. The preference assessment is designed to gather a reasonable amount of judgmental input from the various employee groups. The instrument used for such a task is usually a survey instrument asking questions relative to the concern of the management team. The questions focus on *what is* and *what should be*. The actual development of the questions must utilize a cross-sectional or multidisciplined or cross-functional team of employees.

At this stage of our discussion, one may wonder why so much fuss for something that is—more or less—a world requirement. After all, earlier we emphasized some of the benefits of registration to the ISO standards and it seems reasonable that everyone in any organization should not object to the implementation of such a standard.

So that we may not be misunderstood, let us reiterate that indeed certification will provide marketing opportunities for both existing and future customers. However, if that is the only reason for pursuing certification, the organization will shortchange itself of a greater opportunity to improve.

The focus of the certification ought to be on establishing or improving your quality system. To do that, one must know exactly where to start and how to go about it. Planning for the implementation will start with the needs assessment.

In a typical organization there are three areas where the four phase model will interact with the organization. The three areas are:

1. Organizational analysis
    a. Determine the purpose and parameters of the needs analysis.
    b. Identify and gather information about opportunities for training in the organization.
    c. Identify and gather information about the processes in the organization.
    d. Gather all relevant information.
    e. Analyze the information.
    f. Determine the needs of the organization.
    g. Report the findings.
2. Work behavior analysis
    a. Identify the work to be examined as it relates to specific outcomes.
    b. Construct a plan (flow) for the work behavior analysis.
    c. Job descriptions, job observations, interviews, etc., may be used.
    d. Conduct an analysis of the plan or flow of the work behavior.

e. Gather all relevant information.
f. Determine the needs.
g. Report the findings.
3. Individual capabilities analysis
   a. Identify the characteristics and or capabilities of all employees for specific jobs.
   b. Gather all relevant information.
   c. Analyze and synthesize all data.
   d. Determine the needs of individual employees.
   e. Report the findings to decision makers.

TEAM EFFORT

To carry out an effective needs analysis and the implementation process of the ISO, a team approach must be emphasized. One of the reasons why a team is important is that quality is a collective activity and it transcends both individuals and departments. In fact, for quality to really flourish everyone everywhere in the organization must be committed to both collective and individual responsibility for the benefit of all. It is beyond the scope of this book to discuss the formation of teams, group dynamics, and empowerment. However, for a good discussion of these topics see Scholtes (1988), Jones and McBride (1990), Shonk (1992), Opper and Fersko-Weiss (1992), Aubrey and Felkins (1988), and Wellins et al. (1991).

From an ISO perspective, a team is imperative. No one individual can complete the documentation and/or the implementation process alone. The tasks are lengthy, technical, specialized, and difficult indeed. It is humanly impossible for one individual to have all the knowledge of all the desired disciplines and experiences to complete all the required tasks. One individual may be the leader (facilitator) but not the entire team.

From the very beginning all work should be undertaken as a team project. A project managed by an appointed management representative is highly recommended. He will be responsible not only for overseeing the development and documenting stages, but also for maintenance of the implemented system. A typical representative for such a task is often the quality manager with wide responsibility going outside his own department.

The project manager must define to all team members what the requirements are, what is involved, what the individual team responsibilities are, and what is the time for completion. The project manager is responsible for accurate and timely data. As such, one of the responsibilities that he has to carry out is to make sure appropriate reviews are conducted when applicable with the team members.

THE ISO 9000 IMPLEMENTATION: A PROJECT MANAGEMENT APPROACH

Work—no worthwhile thing has ever been accomplished without it. Work can be either routine and dull or rewarding and energizing. Global competition requires that our attitudes and values toward work be dynamic, engaging, and collaborative. The ISO 9000 standard series is one way to revitalize people, their attitudes, and their approach to work. It is indeed the core minimum of all quality systems taking into consideration the entire organization. An overview of this relationship can be seen in Figure 5.1. Note that in this figure the ISO is not only the center of all functional departments in a given organization—the base of the cone—but cuts through all the levels of personnel (both management and nonmanagement).

There are only a few surprising basic principles associated with ISO, but actually working them into the fundamental work business may be the toughest job of a management team. If these basic principles are allowed to develop, they may change the very nature of work itself.

To focus on this work with the specific goal of *improvement*, at least five issues must be considered, and all are based on a sound needs assessment. They are:

Establish important goals and objectives. The management must define *why ISO* and, perhaps more important, management must define *what is reality* from the current organization's point of view. Is management pursuing ISO because of others or because of a general philosophy to be the best?

Formulate actions via policies, programs, and procedures to achieve the desired goals. Management must be committed rather than involved

Figure 5.1 The relationship between the organization, its human resources, and the ISO structure.

in whatever they say and do. The mode of operation should be to lead *by example* rather than to lead *by memo* or *directive*. It is imperative that these policies, programs, procedures, and goals are based on being purpose-led (vision), planning-oriented, people-centered, process-focused, and performance-based. ALL effort should be directed at eliminating bias and partial information.

Understand the source(s) of resistance and neutralize them. All work creates stress. All work contributes to resistance. However, once the work is understood, it becomes the status quo. It is the job of management—either directly or indirectly—to make sure that everyone in the organization understands that conflict occurs, and that conflict may be normal and expected. To minimize these uncertainties and conflicts, effective vertical and horizontal communications, tactical, and operational plans can be developed by empowered associates who know where the organization is going and are acutely familiar with the strengths and weakness of both the company and its competitors.

Understand that change must not be forced, it must be managed. All change is difficult. As a consequence, the changes in your organization must be well planned and managed if success in the organization is the goal. To facilitate this, pilot projects and temporary change management structures may be considered, and based on the feedback, appropriate action should be taken on a more permanent basis.

Understand that education and training is a must if the implementation process is to be successful. Everyone in the organization has to have the educational knowledge of the basics of ISO. Not everyone in the organization needs to have both formal and extensive knowledge and experience in ISO. The distinction is important, because if the wrong decision is made waste of resources will occur and the implementation process will fall behind. Management must decide the applicability and appropriateness of *why*, *who*, *how*, *what*, and *where* is going to be part of the education and or part of the training.

How is management going to accomplish all these tasks? We recommend that a project management (PM) be at the core of the implementation strategy. A PM approach will facilitate the change at an optimum level throughout the organization. The kind of flexibility that is needed to ensure accurate information, proper direction, and project focus can only be accomplished through PM. The project manager is the appropriate and perhaps the only properly qualified professional to speed up the cumbersome communication paths of the typical hierarchical structure present in most organizations. The relationship between the organization, ISO, and the PM is shown in Figure 5.2.

How is PM going to help the implementation process? To answer that question, we must first address the issue of what PM is. PM focuses on the project, by definition (Kerzner, 1995). A project, on the other hand, is an undertaking that has a beginning and an end and is carried out to meet established goals within: cost, schedule, quality objectives, and optimum resource allocation.

PM brings together and optimizes (always, the focus is on the allocation of resources) rather than maximizes (going all out at the expense of something else: maximization leads to suboptimization) the resources,

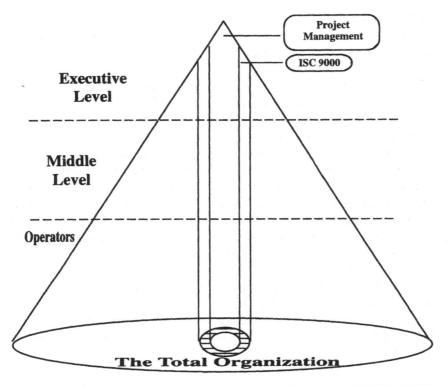

Figure 5.2 The relationship between project management, ISO, and the organization.

which include: skills, talents, cooperative efforts of teams, facilities, tools, information, money, techniques, systems, and equipment.

Since we have defined the implementation process of ISO as a project, it follows that the allocation of the resources in any organization must be important and therefore the appropriate tool for such a task is indeed the PM approach.

Why PM as opposed to other management principles? There are at least two reasons. First, as we have seen, PM focuses on a project with a finite life span, whereas other organizational units expect perpetuity. Second, projects need resources on a part-time and full-time basis, whereas permanent structures require resource utilization on a full-time basis. The sharing of resources may lead to conflict and requires skillful negotiation to see that projects get the necessary resources to meet objectives throughout the project life.

How and what will PM do to facilitate this implementation? PM will follow and assure success of the implementation process by following the four phases of a project's life (Kerzner, 1995). The four phases and the responsibilities of the project manager are given in Table 5.1.

The essence of the ISO 9000 implementation process is that we have defined it as a project. As long as that definition is used in the entire organization, PM is indeed one of the most appropriate ways to pursue the implementation. A typical approach of PM may define the project initiation, understanding the process, the ISO requirements, and the monitoring of the process. An example of the specific items under PM may follow the model shown in Figure 5.3.

Table 5.2 defines a four-phase ISO implementation model with detailed applications in each phase. What is important, however, is that in this model the previous phases may or may not be completed. The implementation process can be parallel (say, two different departments), horizontal (say, within a department), and/or vertical (the entire organi-

Table 5.1 Responsibilities of Project Manager in Implementing Phases of Project Life

Phase	Project manager
1. Defining the project	The primary concern of the project manager is to clarify the project and arrive at an agreement among all concerned about the specific definition and scope as well as the basic strategy for carrying it out. Specific activities may include: • Study, discuss, and analyze the focus of the project. • Write the project definition. • Set an end result objective. • List absolute and desirable needs. • Generate alternatives. • Evaluate alternatives. • Choose a course of action.

Table 5.1 (Continued)

Phase	Project manager
2. Planning the project	Planning means listing in detail what is required to successfully complete the project along the critical dimensions of quality, cost, and time. Specific activities may include: • Establish, review, and modify the project objective. • Choose the strategy for achieving the objective. • Break down the project into small steps. • Determine the performance standards. • Determine realistic time requirements. • Determine the sequence of implementation. • Design a cost budget. • Design the staff organization (internal and/or external). • Determine the appropriate training (internal and/or external). • Develop policies and procedures.
3. Implementing the plan	The entire project is coordinated on an ongoing basis. Specific activities may include: • Control and/or monitor the work. • Negotiate changes. • Provide appropriate feedback (formal and/or informal). • Resolve differences.
4. Completing the project	The goal of PM is to obtain client (in this case, management) acceptance of the project result. This means that management agrees that the quality specifications of the project parameters have been met. For this agreement to take place as smoothly as possible, an objective evaluation must take place based on measurable criteria defined in the early stages of the implementation. (The evaluation may or may not be the ultimate certification through a third-party registrar.) As part of the completion phase it is imperative that steps of follow-up must be defined to make sure that the ISO system—now in place—will not fade away. The definition of this follow-up may be very specific, so that the continuation of the ISO system will be self-sustained.

Table 5.2   ISO 9000 Implementation Model

Phase 1	Management commitment	Establish an ISO implementation team of one person from each functional area as the steering committee. Train those selected in ISO knowledge
Phase 2	Set up structure	Capture company objectives in ISO format: Mission Goals Focus on continual improvement Policies and procedures Quality management commitment

Phase 3	Implementation	To meet the goal of ISO implementation and or certification, examine internal structure and compare it to ISO:
		Determine the departmental objectives
		Review structure of the organization
		Review job descriptions
		Review current processes
		Review control mechanisms
		Review training requirements
		Review communication methods
		  Reports
		  Meetings
		  Record keeping
		Review all approval processes
		Review risk considerations and how they are addressed
		Review all outputs
		Review all action plans

Phase 4	Registrar	Fulfil registrar requirements Audit Surveillance Corrective action Certification Recertification

zation). The pace is and can be different for different departments as well as organizations.

Once the PM is established and understood by the organization, then and only then, we are ready for the full ISO 9000 implementation process. For details on PM see Kerzner (1995), Michael and Burton (1993), and Haynes (1989). The implementation task of ISO is indeed a process. However, it is not necessarily a linear one, but it follows the four phases in Table 5.3. Each phase defines a certain amount of content and understanding of the process. Based on a thorough needs assessment this task of identification (content and understanding) becomes easy as well as manageable. An overview of this implementation process is given in Figure 5.3. This figure not only identifies the four phases of PM, but in addition, it incorporates all the elements of the implementation process and their relationship to the actual certification process. In the next section, a more complete elaboration of the model is given. We identify each stage as a phase because in each phase there may be more than one item of concern. In fact, quite often there are a series of steps within each phase.

THE MODEL OF ISO IMPLEMENTATION

As already discussed, the model of ISO implementation is based on the PM four phases. The details of each phase follow.

Phase 1. Management Commitment

Management must gain the commitment of the entire organization and develop the appropriate strategy.

Phase 2. Setting Up the Structure

Management must develop the organization and train the employees. This development and training may be done through outside sources or internal sources. The development of the organization deals with culture change and paradigm shifts. The training, on the other hand, is more extensive and requires a great commitment of time allocation and all other resources.

One of the major concerns of management is to identify the vehicle of

Table 5.3 The ISO 9000 Implementation Status

Phase 1: Management commitment	Phase 2: Set up the structure	Phase 3: Implementation of procedures	Phase 4: Registrar
Project initiation	Understand process	ISO training	Monitoring progress
Management planning and goal setting	Team flowcharting for process understanding and analysis	Executive training	Worker/operator control in process
Department business and technical commitment	Cause-and-effect analysis	Department training	Define quality manual, procedures, instructions, and forms as they relate to the specific department area
Quality team selected and active	Critical in-process parameters identified	Identification of shortcomings in the system of quality, i.e., specific areas	Internal audit
Training philosophy and tools of quality	Standard operating procedures review, equipment repair, preventive maintenance and calibration	Define boundaries of responsibility	Visit by the registrar
Process definition and selection	Process input and measurement evaluation	Define limitations of resources	Official audit
Critical characteristics identified	Static process data collection	Review system for completeness	Follow-up and maintenance of certification

This grid takes into consideration that the activities in previous phases may or may not be completed. The advantage of implementing ISO 9000 in any organization is that the implementation process can be parallel, horizontal, and/or vertical. The pace and rate of implementation may indeed be different for each individual department. *The long-term goal is to receive and keep the certification.*

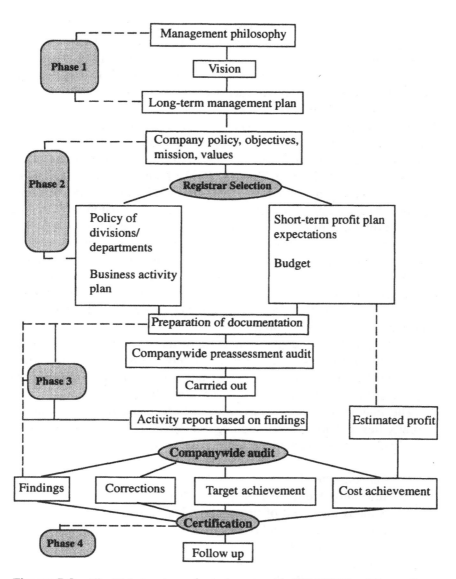

Figure 5.3 The PM structure of a company-wide ISO 9000 implementation model.

transferring the skills, knowledge, and attitudes of a program to the job. The process is called diffusion and it implies sharing with others the knowledge and skills gained during the training. In diffusion the purpose of the sharing is not merely to show (this would be dissemination), but rather to teach. When we teach we give to others the skills, knowledge, and attitudes needed to replicate successes at other training sites.

Since the purpose of diffusion is to teach rather than inform, diffusions themselves must be developed using sound instructional principles. For a detailed analysis of the design of training see Gagne (1977), Gagne and Briggs ((1979), Travers (1982), Fleming and Levie (1979), Briggs (1981), Reigeluth (1983), Wlodkowski (1983), and Stamatis (1986). The training procedures used during the diffusion must ensure that maximum learning (transfer) occurs during the diffusion effort.

Gagne and Briggs (1979) point out that the processes involved in an act of learning are, to a large extent, activated internally. However, these processes may also be influenced by external events, and this is what makes instruction (training) possible. In this sense, instruction is defined as a set of events external to the learner that are designed to support the internal processes of learning.

These external events are a function of management, because management is in fact responsible for the system. As a consequence, management should be able, based on the needs assessment, to identify the proper training and to provide the resources for such training.

The internal events are somewhat different and the responsibility usually lies with the person in charge of designing the training. Gagne and Briggs (1979) have identified nine events:

1. Gaining attention
2. Informing the learner of the objective
3. Stimulating recall of prerequisite learning
4. Presenting the stimulus
5. Providing *learning guidance*
6. Eliciting the performance
7. Providing feedback about performance correctness
8. Assessing the performance
9. Enhancing retention and transfer

The importance of these events in the training environment is that in every event the diffusion should give participants reasons why they should acquire the particular skills and knowledge in the training environment. This is supported by Stamatis (1986), who found that intrinsic motivation and personal attributes do indeed contribute to learning of adults in a technical training environment.

When management or the appropriate representative is ready to begin the training, they should keep in mind four basic characteristics that are unique to adults and should be taken into consideration for the specific training (Kidd, 1973; Fox, 1981; Cross 1982; Stamatis 1986):

1. Adults may differ in their ability to learn. Adults, like youth, vary in their ability to accept new information and adopt new practices. These differences are related to a combination of social, psychological, and physiological factors.
2. Adults may differ in their approach to learning. Adulthood may be characterized by increasing differentiation rather than increasing similarity among individuals. Learning has many functions for adults. The mix of different individuals learning for different purposes makes the approach to learning an important variable in the design of learning experiences for adults.
3. Adults may differ in the ways they process information. Psychological and social factors have a great bearing on the way adults accumulate, store, and retrieve information. The factors at work in the gathering, storing, and retrieving of information have important implications for the practice of adult education and/or training.
4. Adults may differ in the way they think. The way adults acquire content, their cognitive style, and their orientation toward achievement have strong effects on their learning behavior and performance.

How does one go about incorporating these four basic characteristics in the training programs? And what specific training requirements are we interested in for our ISO training? To answer the first question, at least one way, is to follow Carey's (1977) proposal. Carey proposes at least four types of diffusion efforts that can be useful in providing others with the skills and knowledge needed for the job. Briefly, they are:

Type 1 diffusions consist of transportable materials that allow others to learn, practice, and generalize skills from the sponsoring project without the aid of a presenter or facilitator. In this case, the material itself must provide participants with motivation, information, examples, practice opportunities, feedback, generalizations, and applications. Since there is no opportunity for interactive questions and answers, Type 1 diffusions must be carefully prepared, field-tested, and revised, when necessary, before distribution to interested parties.

Type 2 diffusions include transportable materials, plus verbal explanations, examples, and interactive discussions by an expert staff member or members from the sponsoring party. In Type 2 diffusions, the presenter and the materials share the responsibility for providing the participants with motivation, information, examples, and ideas for generalizing to new projects. Practice and feedback activities are still considered a part of the prepared diffusion materials. This type of diffusion is often needed by a new project during the planning activities at the outset of the project.

Type 3 diffusions include material and presenters. The added feature of Type 3 diffusions is the interactive, workshop-like practice and feedback activities. Benefits of such activities include:

- Information from the interaction with the presenter and other workshop participants
- The motivation created by being in a new environment and setting
- The opportunity to practice newly acquired skills in interesting simulations and games
- The opportunity to discuss implementation of new skills in one's own project with others facing similar constraints and decisions

Type 4 diffusions offer all those components mentioned in a Type 3 diffusion with the added characteristic that they are often conducted at a host project site where participants have the opportunity to discuss implementation ideas with project members currently working in the area. A summary of the four diffusion phases is shown in Table 5.4 according to content.

Table 5.4 The Four Diffusion Phases

Type 1	Type 2	Type 3	Type 4
5. Materials alone	1. Materials 2. Presenter	1. Materials 2. Presenter 3. Group interaction	1. Materials 2. Presenter 3. Group interaction 4. Site visit 5. Technical assistance

To address the issue of ISO specific training, the best we can do is to recommend a generic training. Each organization is unique and has specific requirements. However, a typical internal training curriculum for ISO in any company is the following:

Executive overview: This one-day training session should be designed as an introduction to the ISO for executives and senior management. Its function is to educate, remove the mystery and misinformation from the subject, and help the executive team make informed decisions about the benefits, commitment, and plans required to obtain registration. General topics covered are:

- What is ISO 9000?
- Why do you need it?
- How is it organized?
- How do you get it?

Twenty clauses of ISO 9001: This one-day training session should be designed to address the 20 clauses in detail. The focus should be on intent, meaning, and explanation. The audience for this training should be anyone in the organization who wants to learn more about the ISO standard. General topics covered are:

- Understanding the entire ISO 9000 series
- A detailed review of the contents of the ISO 9001
- Applying the understanding of the standards to simulated business situations
- Applying the ISO standard to your organization

ISO 9000 implementation: This 2–3-day training session should focus on a detailed action plan for successful implementation of a quality system. The target audience should be the middle managers, the supervisors, and the personnel who will actually do the documentation. The length of the training depends on the background of the participants and their knowledge of documentation. General topics covered are:

- ISO overview
- A detailed explanation of the ISO 9001
- A general review of ISO 9004
- The implementation strategy for an organization:

    Organize the effort
    Establish a management committee
    Document the system
    Provide for training
    Registration strategy
    Register your company

Documentation: This 2-day training session should be an in-depth review of ISO 9000 requirements and methods for documentation. It should emphasize the different levels of documentation, how to keep documentation simple and effective, and how to maintain control of documentation. The audience for this training should be the supervisor, operators, and anyone who is involved with documentation at any level or wants to understand the ISO documentation. General topics covered are:

- Overview of the ISO 9000 series
- The importance of documentation
- Standard requirements
- Effective documentation
- Format and production issues of documentation
- Document control

Internal auditor: This 2-day specialized training session should cover the concept of audit, the setup and maintenance of an effective audit program, and the difference between inspection and audit. The audience for this training should be a select group of employees who are going to be the designated internal auditors for the organization. General topics covered are:

- General concepts of quality
- Auditing to a standard
- The role of the auditor
- The audit
- The audit administration
- Ethics of audits

Miscellaneous training: Depending on the needs of a specific organization, the following additional training may be necessary:

- Needs assessment
- Project management
- Team building
- Lead auditor training
- Facilitator training
- Problem-solving techniques
- Flow charting

Consulting services: A given organization may require specific consulting services in any area. If additional services are required, a good rule of thumb to follow is to do the training or provide the service on a *just-in-time* basis. Typical services are:

- Specific training
- Readiness review
- Documentation assistance
- Specialized implementation services
- Supplier evaluation

Phase 3. Implementing Procedures and Documenting the Quality System

In this phase the structure of the documentation is defined and developed. Specifically, the following items are considered:

1. Identify all pertinent procedures, policies, and practices to meet ISO 9000 and/or customer-specific requirements.
2. Prepare the documentation. The hierarchy of the tiers is shown in Figure 5.4. For an overview discussion of documentation, see Chapter 6.
   a. Quality manual. This is a roadmap to the system, outlining the policies and objectives that relate to specific aspects of the system.
   b. Procedures. These provide process descriptions and flow-charts of activities. They give more detail of what, who, where, why, and when an activity is carried out.
   c. Work instructions. These describe step by step how to carry out a task. They are often called *standard operating procedures* (SOP) or *standard job practices* (SJP) or *operating guides*. Work instructions must be revised and integrated into the overall documentation system.

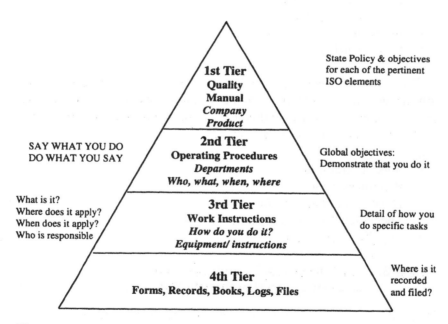

Figure 5.4   The four hierarchical tiers of documentation.

d. Forms and records. Forms are often used to collect information and record the completion of required quality activities. Sufficient records must be kept to provide objective evidence that the quality activities are being carried out.

Phase 4. Working with the Registrar

This phase provides for the closure in the implementation process and is pretty much controlled by the registrar. The contribution of your company is minimum at this stage, although very important. Issues of cooperation, open communication, true expectations are some of the concerns of this phase. Specific characteristics of this phase are: pre-assessment, site visit (audit), registration or corrective action, and follow-up.

Finally the overall structure of a company-wide ISO 9000 implementation may be seen in Figure 5.5. This figure not only identifies the individual phases that the implementation process must go through, but in addition it identifies the sequence of events that must be followed in the pursuit of certification.

One of the most important steps in this figure is identification of the registrar selection during phase 2. This is very important since it may take some time to secure a registrar for the certification process. Be prepared as early as possible. Just because you are ready, it does not mean that the registrar can schedule you right away. It has been reported that as of this writing, some registrars have waiting lists as long as 12 months.

Since the ISO structure provides for the foundation of quality, it is by definition less stringent than the total philosophy of TQM. To be sure, contrary to public opinion the ISO provides for *continual improvement* in the scope of the standard, the corrective action subsection, and elsewhere. To move from ISO to TQM requires an extension of all the basic quality guidelines as found in the ISO and an active initiative in the principles of:

Measurement
Customer satisfaction through active listening for their needs, wants, and expectations
Continual adaptation to changing market conditions

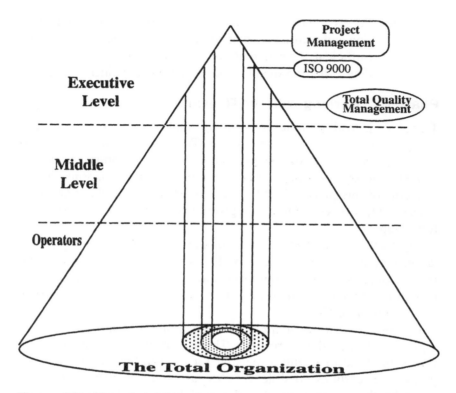

Figure 5.5 The relationship of project management, ISO, TQM, and the organization.

    The best is yet to come
    Market-driven attitude

The expanded relationship of the ISO and TQM can be seen in Figure 5.5. For more information on this relationship, see Chapter 10.

COST OF IMPLEMENTATION

The discussion of implementation always raises concern about *how much it is going to cost* for a program that is so heavy in documentation.

Although nobody can really identify individual costs for individual companies, there are some general guidelines. For example, we already know that the majority of the cost is an internal cost rather than a cash disbursement to outsiders. Most of the cost is associated with writing manuals, procedures, instruction, etc. The amount of these costs obviously depends on where the organization is in relationship to the requirements.

Cost is related to effort and time. Therefore, depending on how much time and effort your organization is willing to invest for the total implementation, the cost is going to reflect both. It is known, however, that for organizations with some or substantial documentation in place, the cost is somewhat small. On the other hand, for organizations with no documentation, the cost is quite large. In Chapter 7 we will discuss the issue of effort and time and cost in more detail.

Table 5.5 Estimated Implementation Costs

Area	Low cost	High cost
Consulting	$ 20,000	$ 50,000
20–30 days		
Training	5,000	20,000
Public seminars		
Software tools	—	20,000
QA applications		
Word processing		
Registrar		
Preassessment	5,000	10,000
Assessment	10,000	45,000
Annual surveillance	5,000	15,000
Subtotal	$ 45,000	$160,000
Internal costs	100,000	150,000
Meetings		
Writing manuals		
Reviews		
Visits		
(approximate time 10–24 months)		
Grand total	$145,000	$310,000
(average)	$227,500	

Are there any guidelines for cost? In a special report in July 1993, the *Continuous Improvement* newsletter published the survey results from a national study on ISO consultants. That study found that there were 1800 ISO consultants charging anywhere from $320 per day to $2000 per day with an average fee of $958.36 per day.

Table 5.5 gives estimated costs based on the author's experience (having participated in over 20 implementations and registrations for both ISO 9001 and 9002 standards).

A SUMMARY OF THE ISO 9000 IMPLEMENTATION

The needs assessment will be the first item of action in the process of ISO implementation. The needs assessment will be the *engine* and the *driver* for the specific requirements of your organization. Once the needs assessment has been completed, then the following may be addressed. (Note that the outline may be too prescriptive for your needs. Remember, this is only a guideline and for some organizations some of the items may not apply. On the other hand, for some organizations it may not be enough.)

1. Scope of the certification
   a. Location
   b. Business unit
   c. Product line
   d. Appropriate standard
2. Establishing the implementation team
   a. Project leader
   b. Core team
   c. Functional representatives
3. Education and training
   a. Public seminars
   b. Lead auditor training
   c. Orientation programs
   d. Documentation methods
   e. Internal auditor training
   f. Employee update meetings

      g.   Networking
4.   Effective use of consulting resources
      a.   Primary benefits
      b.   Suggestions
      c.   Selection criteria
5.   Documentation
      a.   Generic structure
      b.   Alternative structures
      c.   Standard format
      d.   Software tools
6.   Operating the quality system
      a.   Document control
      b.   Internal auditing
      c.   Corrective action
      d.   Management review
7.   Registrar selection
      a.   Services
      b.   Selection criteria
8.   Third-party assessments
      a.   Preassessment
      b.   Initial assessment
      c.   Surveillance audits

REFERENCES

Aubrey, C. A., and Felkins, P. K. (1988). *Teamwork: Involving People in Quality and Productivity Improvement*. Quality Press, Milwaukee, WI.

Briggs, L. J., Ed. (1981). *Instructional Design*. Educational Technology Publications, Englewood Cliffs, NJ.

Carey, L. M. (1977). "Diffusion." In Keith A. Acheson, Ed. *The Five Dimensions of Demonstration*. Teachers Corps Research Adaptation Cluster, University of Oklahoma, Norman, OK.

Cross, K. P. (1982). *Adults as Learners*. Jossey-Bass, Washington, D.C.

Fleming, M., and Levie, H. (1979). *Instructional Message Design: Principles from the Behavioral Sciences*. Educational Technology Publications, Englewood Cliffs, NJ.

Fox, R. (1981). *Current Action Principles and Concepts from Research and*

Theory in Adult Learning and Development. ERIC Document Reproduction Service No. ED. 203 008.

Gagne, R. M. (1977). *The Conditions of Learning*, 3rd ed. Holt, Rinehart & Winston, New York.

Gagne, R. M., and Briggs, L. J. (1979). *Principles of Instructional Design.* Holt, Rinehart & Winston, New York.

Haynes, M. E. (1989). *Project Management: Four Steps to Success.* Crisp Publications, Los Altos, CA.

Jones, L., and McBride, R. (1990). *An Introduction to Team-Approach Problem Solving.* Quality Press, Milwaukee, WI.

Kaufman, R., and English, F. W. (1979). *Needs Assessment.* Educational Technology Publications, Englewood Cliffs, NJ.

Kerzner, H. (1995). *Project Management: A Systems Approach to Planning, Scheduling and Controlling*, 5th ed. Van Nostrand Reinhold, New York.

Kidd, J. R. (1973). *How Adults Learn.* Association Press, New York.

Michael, N., and Burton, C. (1993). *Basic Project Management.* Singapore Institute of Management, Singapore.

Opper, S., and Fersko-Weiss, H. (1992). *Technology for Teams.* Van Nostrand Reinhold, New York.

Reigeluth, C. M., Ed. (1983). *Instructional Design Theories and Models: An Overview of Their Current Status.* Lawrence Erlbaum Associates, Hillsdale, NJ.

Scholtes, P. R. (1988). *The Team Handbook.* Joiner Associates, Milwaukee, WI.

Shonk, J. S. (1992). *Team Based Organizations: Developing a Successful Team Environment.* Quality Press, Milwaukee, WI.

Stamatis, D. H. (1986). *The effects of hierarchical and elaboration sequencing on achievement in an adult technical training program.* Unpublished dissertation. Wayne State University, Detroit, MI.

Travers, R. M. W. (1982). *Essentials of Learning: The New Cognitive Learning for Students of Education.* Macmillan, New York.

Wellins, R. S., ByHam, W. C., and Wilson, J. M. (1991). *Empowered Teams.* Quality Press, Milwaukee, WI.

Wlodkowski, R. J. (1985). *Enhancing Adult Motivation to Learn.* Jossey-Bass, Washington, D.C.

6
Documentation Overview

This chapter addresses—from an overview perspective—the development and ongoing maintenance of

The documentation of standards and procedures
The documentation of records from the results of inspections, reviews, and audits

As already discussed, the ISO standards enable firms involved in international trade to obtain a degree of confidence in the quality of the work done by current and/or potential suppliers. Furthermore, the assumption on which they operate is that if the process is effective, the product and/or service will, more likely than not, be of high quality.

This effectiveness, however, involves compliance with a set of standards and procedures, which the organization has defined based on its products or services. Compliance, on the other hand, is assured through a formal system of inspection and audits.

ISO GUIDELINES FOR DOCUMENTATION

This formality is addressed in the ISO 9000 as a requirement that a firm's processes be documented, and the firm conform to the statements of its own process.

All the certifiable standards (ISO 9001, ISO 9002, ISO 9003) require that appropriate and applicable documentation be in place. This presents a major problem since unless something is in writing, it cannot be documented. In fact, the problem is so great that the most common reason for failure during registration is related to inadequate documentation (CEEM, 1993; National ISO 9000 Support Group, 1993).

To put our confusion to rest, ISO 9004, clause 5.3.1 provides a very concrete statement:

> All the elements, requirements, and provisions adopted by an organization for its quality management system should be documented in a systematic, orderly, and understandable manner in the form of policies and procedures.
>
> The quality system should include adequate provision for the proper identification, distribution, collection, and maintenance of all quality documents.

To satisfy this ISO guideline there are three requirements:

Requirement 1:   Documentation of Specific Quality Standards and Procedures

Requirement 1 is perhaps the easiest to comply with, since the company defines *how they do what they do*. The intent here is to ensure that a firm's way of doing business can be evaluated and it follows the ISO standard. The organization is expected to produce a true reflection of the way the firm does business and the way employees and suppliers perform their functions.

This fundamental principle of standard compliance can be summarized in:

Say what you do,
Do what you say you do

and formalized in the quality manual, quality plans, and procedures.

Requirement 2: Documentation of Product Development (Design and Development) Life Cycle Results

Requirement 2 consists of specifications, design documentation, test plans, and other descriptions of the product or the development and acceptance process. Its purpose is to establish continuity and control during product development.

Requirement 3: Documentation of Outcomes Required Under the Standards

Requirement 3 represents the verification of outcomes from specific tasks, reviews, audits, inspections, etc. These are the quality records, and they produce an audit trail that is used to verify that the organization performs the functions described in its standards and procedures documentation. Quality records, in general, provide the basis for performance evaluation toward continuous process improvement.

So far, we have addressed the issue of documentation from the standard's perspective. Let us now look at some of the specific benefits that the organization may gain.

First, all documentation—regardless of source—should fulfill a purpose that should benefit the organization in some way either directly (internally) or indirectly (externally).

Second, all documentation should be as specific as possible where appropriate. It is this specificity that will guide the organization in continual improvement. For specific benefits, let us examine the following:

Standards and Procedures

If standards and procedures indeed represent reality (current practice) and are accurate, they will contribute to the firm in ways that:

- Make the training of new staff easier
- Permit more objective evaluation of performance
- provide continuity and consistency when there is turnover of staff
- Facilitate training of specific tasks for current employees
- Provide a benchmark for future improvement

Quality Records

Quality records provide the evidence and documentation of compliance with preestablished standards and procedures. In addition, as records accumulate, they will provide a data base for performance analysis and performance improvement.

Life Cycle Documentation

Documentation provides continuity across product development life cycle, which allows for project control and product quality evaluation during the development process. It also helps in future development of similar products.

Document Control

Document control is a requirement. Document control must be accessible to those who have need and authorization to use it and it must be kept up to date.

STRUCTURE OF THE DOCUMENTS

The structure of documentation is as important as its content. The structure is the way the information is organized. It enables the reader to find information easily and to access only what is of interest.

Structure can be viewed from two points of view: level of detail within subject areas and parsing, based on content, into subject areas. Both points of view are legitimate and appropriate. Their use is dependent on the organization and the writer.

If the documentation is expected to be procedural and/or technical, a hierarchical structure is effective and highly recommended. In a hierarchical structure, the information is presented in levels of detail. The hierarchical structure starts with an overview, followed by one or more levels of detail. A typical hierarchical structure is shown in Figure 6.1.

The overview of the hierarchical structure gives the reader the whole idea of a topic. In essence, the overview is the roadmap of the detailed information and should allow the reader to select the specific part of the topic that is of interest or to skip the topic entirely. An example of an overview in procedures may be a process flow diagram or the objective of the procedure.

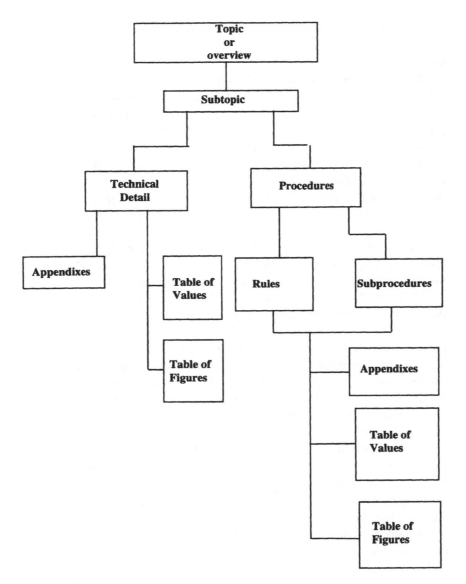

Figure 6.1 A typical hierarchical content design structure.

In the area of documentation overviews are encouraged and their frequency or length depends on the complexity and/or levels of the specific documentation.

The first level of detail is usually the material that provides the main substance of the topic to the reader. In a procedure, it describes the specific steps to be taken so that the procedure's objective is accomplished. When the levels of detail become cumbersome, they should be put in an appendix or some other reference area.

With the proliferation of computer technology, documentation structure can be fun and easy to design. The idea of a computer-generated documentation is that it will allow you to either *chain* or *jump around* in your documentation. The facilitation of chaining and or jumping around will depend on the software and the ability of the user. Some commercial software programs available for such tasks are: Quark Xpress, Hypertext, Pagemaker, WordPerfect. Figure 6.2 shows how the documentation structure may look with the computer design.

While the hierarchical structure defines the level of complexity, parsing is a way of separating things into logical groupings. It is a very subjective activity and care should be taken to make sure that the parsing really reflects common ground of all involved. An example of parsing may be when information is grouped by volumes, chapters, sections, etc. In procedure writing, a given procedure may be parsed based on the type of work, the performer, the time in the process, or alphabetically by the title of the procedure. A pictorial view of parsing is shown in Figure 6.3.

A volume is a bound document (paper document), generally no more than one and a half inches thick. It is defined by the user, the time, and the frequency of use as well as the level of detail. Whenever possible, use only one volume. Volumes may be collected and bound so that they are kept together.

An appendix is a stand-alone (supplemental rather than essential) document that adds to the main information in the chapters, sections, and/or subsections.

Chapters, sections, and subsections are based strictly on the logic of the material. The deciding factor should be the need and convenience of the reader. A chapter is a combination of sections.

A section is a labeled set of paragraphs, graphics, or other elements that fully address the topic. A section is a combination of subsections.

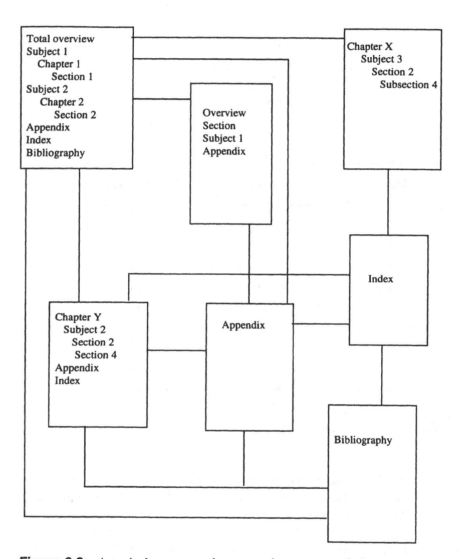

Figure 6.2 A typical computer documentation structure design.

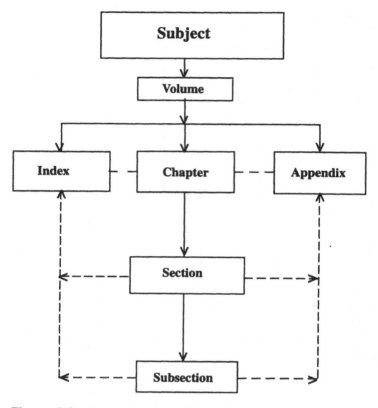

Figure 6.3   A typical content hierarchy.

A subsection is one to three paragraphs that address a specific topic.

An informational element is the finest level of detail that expresses a complete piece of information relevant to the topic. Usually it is found as a definition, a step procedure, a description of a part/component, or sometimes an overview of a topic. An informational element is a subsection at the lowest level of the hierarchy.

The discussion of hierarchical structures and parsing may be frightening to someone who is exposed for the first time to any form of documentation. However, there are some basic rules that define the specific breakdown of a particular project.

1.   There is no *right* approach for the documentation structure. One

may use either the hierarchy or the parsing method. The selection process is highly subjective and a matter of preference. For example, for procedural documentation, the flow of the actual procedure itself is the most common basis of the breakdown.

2. The hierarchy and parsing in documentation are used only to make the writing easier because the writer can divide the document into small parts and write each one independently. In fact, it is this attribute that allows several people to write parts of the same document.

In addition, the reading becomes easier because the reader is advised of the content of a section before reading it. This facilitates comprehension.

3. When ready to write any form of documentation, plan the writing based on an outline that should establish the contents of the document. Once the writing begins, the outline (the content hierarchy) keeps the writer on track. However, it is possible (sometimes quite often) to change the outline several times during the writing process.

4. When the documentation is complete, a table of contents should be provided. The table of contents permits the reader to find topics and to get an overview of the document contents. Sometimes the table of contents is divided into sections and subsections to make it easy for the reader to find a particular topic. Generally, the table of contents is written after the completion of the document.

5. In any form of documentation, it is strongly suggested that indexing be provided. Indexing is a cross-reference mechanism that provides for locating subjects across the content hierarchy. The structure and the detail of the indexing are the writer's choice. However, the writer should have in mind the need of the cross-reference item and the ability of the reader to find the information. Generally, the index is written after the completion of the entire documentation. However, it is suggested that, for convenience and accuracy, all indexing should be done throughout the writing process.

REFERENCES

National ISO 9000 Support Group (February 1993). "How to prepare your documentation for ISO 9000 registration." *Continuous Improvement Newsletter*. National ISO 9000 Support Group, Caledonia, MI.

CEEM. (September 1993). *ISO 9000 Survey. Quality Systems Update*. CEEM Information Services, Fairfax, VA.

7

ISO and Services

This chapter addresses the issues of the ISO quality standards and the service industry. Our aim is to show that the structure of the ISO 9001 is indeed applicable to the service industry, and it provides the building blocks of a quality system for all service organization. The implementation of the ISO 9001, however, is enhanced with the ISO 9004-2 guideline.

OVERVIEW OF SERVICE

Schwartz (1992) and Stamatis (1995) have differentiated manufacturing and service in a variety of ways including:

The majority of goods are:	The majority of services are:
Tangible	Intangible
Storable	Perishable
Not for immediate consumption	Immediate consumption

The intangibility of services makes it much more difficult to apply skills of good quality control to service. These skills include, but are not limited to, the following:

Determining and specifying quality of design
Setting standards
Measuring for performance
Rejecting defects
Regulating the process ((service)
Utilizing feedback
Redesigning the service

The concern, then, that most service managers have in reference to ISO 9000 is: how does the standard apply and how is customer satisfaction accounted for?

To be sure, quality and customer satisfaction in service are important subjects and are receiving increasing attention worldwide. However, since the ISO 9001 does not really address the issues of service, the International Organization for Standardization developed a subsection to the ISO 9004 guideline to provide a response to this awareness. The guideline, called ISO 9004-2, seeks to encourage organizations to manage the quality aspects in a more consistent and effective manner. The ISO 9004-2 builds on the principles given in the ISO 9000–9004 series. It recognizes that a failure to meet quality objectives can have consequences that may affect the customer, the organization, and society. Furthermore, the standard recognizes that it is management's responsibility to ensure that such failures are prevented.

The successful application of service quality management provides significant opportunities for improved service performance and customer satisfaction; improved productivity, efficiency, and cost reduction; and improved market share.

To achieve these benefits, a quality system for services should also respond to the human aspects involved in the provision of a service by:

- Managing the social processes involved in a service
- Regarding human interactions as a crucial part of service quality
- Recognizing the importance of a customer's perception of the organization's image, culture, and performance

- Developing the skills and capability of personnel
- Motivating personnel to improve quality and to meet customer expectation

OVERVIEW OF THE GUIDELINES

It is important to recognize that the ISO 9004-2 is not a certifiable standard. Rather, it is a set of guidelines for all services to follow in their quality quest for continual improvement. The guidelines provide the basic blocks of quality for all services and encourage them to focus on customer satisfaction.

The guidelines provide a systematic path of defining service, delivery, and *how* the organization can go about improving the quality system in the short and long term.

All services are included, if they are indeed interested in quality. All services can improve, if they are indeed focused on the customers and their needs. The ISO structure provides the path for that improvement. Types of service industries that can be active participants in the ISO are:

Hospitality	Financial
Health	Communications
Professional	Administration
Maintenance	Technical
Purchasing	Scientific
Trading	Utilities

QUALITY MANAGEMENT AND QUALITY SYSTEM ELEMENTS

The aim of this section is to summarize each of the clauses in the guidelines and to provide an overview of what service quality is all about from the ISO perspective.

Section 1: Scope. The guideline establishes the need for implementing a quality system in the service industry.

Section 2: Normative References. The guideline provides for the references that are used to define the service quality.

Section 3: Definitions. In this section the specific vocabulary for
the service is utilized and is based on the ISO 8402.

Section 4: Characteristics of Services. In this section the guideline
provides for an overview of what the service and service delivery
characteristics are all about. Furthermore, this section defines the control of services and service delivery characteristics.

Section 5: Quality System Principles. This section delineates the
key aspects of the quality system and identifies each of the components
as follows:

- Management responsibility
- Quality system structure
- Personnel and material resources
- Interface with customers

Section 6: Quality System Operational Elements

Section 6.1: Marketing Process

Section 6.1.1: Quality in Market Research and Analysis. This
clause defines the responsibility of marketing and promotes the need
and demand for a service. The clause also expects the management to
establish procedures for planning and implementing market activities.
Specifically, it identifies the following as associated elements:

- Establishment of customer needs
- Complementary services
- Competitor analysis
- Review of appropriate legislation and standards
- Analysis and review of customer requirements
- Consultation with all affected functions to confirm the commit-
  ment and ability to meet service quality
- Ongoing research to examine changing market needs
- Appropriate application of quality assurance/control

Section 6.1.2: Supplier Obligations. This clause requires that the
supplier obligations to customers must be noted and communi-

cated to the service organization and the customer. The obligations should be consistent with:

- Related quality documentation
- Supplier capability
- Relevant regulatory requirements

Section 6.1.3: Service Brief. This clause reiterates the need for a definition of the customer's needs and the capability of the organization to provide the identified need.

Section 6.1.4: Service Management. This clause defines the responsibility of management in regard to:

- Development and withdrawal of the service
- Ensuring appropriate resource allocation
- Planning

Section 6.1.5: Quality in Advertising. This clause reflects the requirements of *customer's perception* as a service specification and it recognizes the liability risks of offering exaggerated or unsubstantiated claims for the service.

Section 6.2: Design Process

Section 6.2.1: General. This clause delineates the concept of service brief. Specifically, it defines the service to be provided, delivered, and controlled.

Section 6.2.2: Design Responsibilities. This clause defines specific responsibilities to management such as:

- Planning, preparation, validation, maintenance, and control
- Specifying products and services to be produced
- Implementing design reviews for each phase of the service design
- Validating that the service delivery process meets the requirements
- Updating the service specification

In addition to the design responsibilities, it is important to recognize the service, quality, and delivery specifications during the design. Specific considerations may include:

- Planning for variations
- Carrying out a "what if" scenario with both systematic and random failures
- Developing contingency plans

Section 6.2.3: Service Specification. This clause identifies the need for clarity in the description of the service and allows for a standardized acceptability for each service characteristic.

Section 6.2.4: Service Delivery Specification

Section 6.2.4.1: General. This clause defines the requirements of delivery. Specifically, it identifies guidelines for optimum results, such as:

- A clear description
- A standard of acceptability
- Resource requirements
- Number of skills of personnel required
- Reliance on subsubcontractors

Needless to say, these guidelines should follow both the delivery specifications and the legal requirements for health, safety, and environmental protection.

Section 6.2.4.2: Service Delivery Procedures. This clause provides for flexibility in the service to subdivide the process into separate work phases.

Section 6.2.4.3: Quality Procurement. This section provides strong guidelines for the requirements of procured products and services. Some considerations are:

- Provision for descriptions and/or specifications in purchase orders
- Selection of qualified subcontractor

- Agreement on quality requirements
- Provision for disputes
- Incoming product and service controls
- Incoming product and service quality records

In addition, this clause provides some guidelines for selecting the subcontractor by having:

- On-site assessment and evaluation of the subcontractor
- Evaluation of subcontractor samples
- Past history of subcontractor
- Test results of similar subcontractors
- Experience of other users

Section 6.2.4.4: Supplier-Provided Equipment to Customers for Service and Service Delivery. This clause is a reminder that when supplier equipment is provided, it must be suitable for its purpose and that written instructions must be given, as required, for its use.

Section 6.2.4.5: Service Identification and Traceability. This clause emphasizes the awesome responsibility of identification and traceability.

Section 6.2.4.6: Handling, Storage, Packaging, Delivery, and Protection of Customer's Possessions. This clause reiterates the need for effective controls for the handling, storage, packaging, delivery and protection of customer's possessions that the service organization is responsible for.

Section 6.2.5: Quality Control Specification. This clause defines the scope of quality in service and provides specific guidelines for the appropriate design of quality control specification.

Section 6.2.6: Design Review. This clause reiterates the need for design review against the service brief. The design review should be consistent with and can satisfy the requirements of:

- Items in the service specification
- Items in the service delivery
- Items in the quality control

Participants at each design review should include representatives of all the functions affecting service quality. The design review should anticipate problem areas, identify inadequacies, and initiate actions to ensure that:

- All specifications meet the customer's requirements
- The quality control specification is adequate to provide accurate information about the quality of service delivered

Section 6.2.7:   Validation of the Service, Service Delivery, and Quality Control Specifications.   This clause defines when and how validation should be carried out and under what conditions. In addition, it recommends that the validation should confirm that:

- The service is consistent with customer requirements
- The service delivery process is complete
- Resources are available to meet the service obligations
- All applicable codes, practices, and standards are satisfied
- Information is available to customers on the use of the service

Section 6.2.8:   Design Change Control.   This clause not only defines the objective of design change control, but recommends several points for its insurance. The points are:

- Know the need for change
- Changes are planned
- Appropriate personnel are ALWAYS participative on the functions affected by the change
- Customers must be notified when changes are affecting them
- The impact of change should be evaluated

Section 6.3:   Service Delivery Process

Section 6.3.1:   General.   This clause reiterates the need for management to assign specific responsibilities to all personnel implementing the service delivery process.

Section 6.3.2: Supplier's Assessment of Service Quality. This clause emphasizes the need for quality to be part of the operation of the service delivery process. This includes:

- Measurement and verification
- Self-inspection
- A final supplier assessment

Section 6.3.3: Customer's Assessment of Service Quality. This clause reiterates the importance of the customer's assessment and reassures us that unless we take measures to identify and measure the quality of a service, we may end up with "some" customer reaction.

Section 6.3.4: Service Status. All work done should be recorded to identify the achievement of the service specification and customer satisfaction.

Section 6.3.5: Corrective Action for Nonconforming Services

Section 6.3.5.1: Responsibilities. This clause puts the responsibility on each individual to identify and report ALL nonconforming services. Corrective action should be defined appropriately.

Section 6.3.5.2: Identification of Nonconformity and Corrective Action. This clause recommends that nonconformities, when detected, should be recorded and action should be taken to analyze and correct them.

Section 6.3.6: Measurement System Control. This clause emphasizes the need for establishing, monitoring, and maintaining procedures for service measurement.

Section 6.4: Service Performance Analysis and Improvement

Section 6.4.1: General. This clause discusses the need for continual evaluation of the operation of the service processes, which should be practiced to identify and actively pursue opportunities for service quality improvement.

Section 6.4.2:   Data Collection and Analysis.   This clause emphasizes that the decisions of the service organization should be based on data from the service operation.

Section 6.4.3:   Statistical Methods.   The use of statistical methods is encouraged in this clause, as well as processes in control. However, it is up to the individual company, where applicable and appropriate, to prescribe the optimum statistical methods.

Section 6.4.4:   Service Quality Improvement.   This clause recommends the establishment of programs that support the continual improvement process in the service quality, including an effort to identify:

- The most important characteristics to the customer
- Any changing market needs
- Any deviations from a specific service
- Opportunities for reducing cost

The activities recommended in this clause address the need for both short-term and long-term improvements and include:

- Identifying relevant data
- Data analysis
- Reporting periodically to senior management

IMPLEMENTATION OF THE ISO STANDARDS IN THE SERVICE INDUSTRY

The implementation process for the ISO standards in any service industry is exactly the same as in any other organization. The steps of implementation are the same as those identified in Chapter 5.

REFERENCES

ISO 9001:1994. *Quality Systems—Model for Quality Assurance in Design/ Development, Production, Installation and Servicing.*
ISO 9002:1994. *Quality Systems—Model for Quality Assurance in Production and Installation.*

ISO 9003:1994. *Quality Systems—Model for Quality Assurance in Final Inspection and Test.*

ISO 99004-2: 1991(E). *Quality Management and Quality System Elements. Part 2: Guidelines for Services.*

ISO 10011-1:1990. *Guidelines for Auditing Quality Systems. Part 1: Auditing.*

ISO 10011-2:1991. *Guidelines for Auditing Quality Systems. Part 2: Qualification Criteria for Quality Systems Auditors.*

ISO 10011-3:,1991. *Guidelines for Auditing Quality Systems. Part 3: Management of Audit Programs.*

ISO 10012-1:1991. *Quality Assurance Requirements for Measuring Equipment. Part 1: Management of Measuring Equipment.*

Schwartz, M. H. (1992). "A question of semantics." *Quality Progress.* ASQC, Milwaukee, WI.

Stamatis, D. H. (1995). *The Principles of Total Quality Service.* St. Lucie Press, Delray, FL.

8

ISO and Software

This chapter addresses the issues of the ISO quality standards and the software industry. Our focus is the software industry and how it is affected by the international standards. In our discussion we will give a background of the software industry and we will summarize both the current activity and the ISO 9000-3 as they apply to the industry at large. In our summary of the ISO 9000-3 we will show that indeed *all* software services can and should follow the building blocks of quality as identified by the ISO structure.

OVERVIEW

Software development is a relatively new discipline. There is still no consensus on its body of knowledge and how to produce consistently good results. It is notoriously difficult to assure that software works correctly or as it was originally intended. Even today, it is still not practical to test software 100% completely to assure that it complies with all requirements. Those who have worked in software or been involved as users, suppliers, customers, or managers know from experience that there is a lot of risk in achieving necessary usability, cost, accuracy, timeliness, maintainability, expendability, and other requirements. Although there are many efforts to turn software development from an art form into a true engineering discipline, this is still largely an immature field.

Many IEEE Computer Society standards have been developed to assist in making software development more engineering-like. The Software Engineering Institute has developed the capability maturity model (CMM), which is beginning to be widely used. In the last few years an international standards subcommittee in software engineering has taken on a number of new standards projects, including the further development of CMM in the international arena. It is now recognized that software has many intrinsic differences that make it difficult to standardize, predict in terms of project size, time, and resources, measure, evaluate, and manage. We know that although specialized training, experience, and skills are required for success, they do not guarantee success. The tools, methodologies, and systems just are not available yet.

Unlike many other disciplines, most of the work in software development is largely intellectual and consists of getting the requirements right, developing the system architecture, and designing and developing detailed programs. The emphasis in terms of ISO 9001 is on design, not "production." Therefore, the design control element in the software life cycle is especially critical. The production phase, so important in ISO 9001, is quite straightforward and relatively trouble-free in software development. We see, therefore, that ISO 9001 alone does not adequately fit the software development profile. This is why ISO 9000-3 was developed; it adds value because it interprets ISO 9001 for application to software and uses the language of software development.

Quality is being addressed in software development and there are efforts to put in place standards, best practices, and methodologies to assure that quality will be built into software in the early phases of the life cycle. Much work has been done in making the process of writing requirements specifications more explicit, understandable, and representative of end-user requirements. Quality planning aims to assure that quality is built into the product in the early stages and not *tested in* at the end of the development cycle. Prototyping and other automated techniques of describing and demonstrating or simulating user requirements are available and increasingly being used across the software development industry.

Along with these efforts we now have the opportunity to use the ISO 9000 registration process to improve software quality and usability. But to do that requires that the original quality system process as written in ISO 9001 be adapted, since ISO 9001 was written without software in mind. This was the reason ISO 9000-3 was written: to give guidance on how to apply ISO 9001 to software. This is now official ISO guidance, which should be used in quality systems and audits where software is the product or to the extent it affects the quality of the service or the product delivered to the customer.

RELATIONSHIPS AMONG ISO 9001, ISO 10011, AND ISO 9000-3

One of the great advantages of the ISO 9000 series standards is that they are generic and can be applied to any industry anywhere. In the United States currently, the ISO 9000 registration now in place is generic. The United States, for all intents and purposes, does not have any programs for a specific industry—as of September 1, 1994, the automotive industry developed such a program and called it QS 9000 quality system (such programs are called sector programs). Although the system in place in the United States does require that registrars and auditors carry out audits to ISO 10011 using competent auditors, there is no specific requirement for software workplace experience and training. Something is needed to specify qualifications to account for the differences that exist between auditing software development processes and other industrial processes. ISO 9000-3 is used to translate ISO 9001 into the language of computer software developers and users. A comparison between ISO 9001 and ISO 9000-3 is shown in Table 8.1.

In the United Kingdom there is a program (TickIT) that certifies the software companies; in the United States as of this writing, the Software Quality System Registration (SQSR) is proposing a similar program. The recommended computer software program (leading to separate

Table 8.1 Comparison of ISO 9001 and ISO 9000-3

Clause in ISO 9001	Description	Clause in ISO 9000-3
4	Quality system requirement	4, 5, 6
4.1	Management responsibility	4.1
4.2	Quality system	4.2, 5.5
4.3	Contract review	5.2, 5.3
4.4	Design control	5.3, 5.4, 5.5, 5.6, 5.7, 6.1
4.5	Document control	6.1, 6.2
4.6	Purchasing	6.7
4.7	Purchaser supplied product	6.8
4.8	Product identification and traceability	6.1
4.9	Process control	5.6, 6.5, 6.6
4.10	Inspection and testing	5.7, 5.8, 5.9
4.11	Inspection, measuring, and test equipment	5.7, 6.5, 6.6
4.12	Inspection and test status	6.1
4.13	Control of nonconforming product	5.6, 5.7, 5.9, 6.1
4.14	Corrective action	4.4
4.15	Handling, storage, packaging, preservation, and delivery	5.8, 5.9
4.16	Quality records	6.3
4.17	Internal quality audits	4.3
4.18	Training	6.9
4.19	Servicing	5.10
4.20	Statistical techniques	6.4

registrars, separate audits, separate fees, etc.) builds on these standards and places more specific language into the requirements for experience and training regarding software and quality for auditors. There is nothing unreasonable in this requirement—5 years of practical experience in software with 2 of the last 4 years in quality. The chain of requirements is completed by assuring that not only will the auditors have had the training and experience, but that the registrars will use these auditors when providing quality systems registration for a supplier.

Essential Elements in the Software Program

The elements of the SQSR program that are emphasized are the auditor requirements, training and certification, and the accredited registrar's required use of such certified auditors in any audits where the registration would cover the software scope. We will identify all the elements contained in the current *SQSR Guide*. Not all of these elements are of equal importance; for this reason they are grouped into primary and secondary categories. These elements are contained in the Introduction of the *SQSR Guide* under the heading "Uniform Accreditation Arrangements."

Primary

1. Use of ISO 9000-3 as the guide in installing and auditing quality systems where software development is included in the scope of the registration.
2. Registrar's required use of auditors who meet qualification and are certified under the software registration program involving workplace experience in software development and in quality and successful completion of the lead auditor training course containing training on ISO 9000-3 as well as ISO 9001 and case studies based on audits of software development.
3. Registration cycle of 3 years starting with full quality system audit of all 20 elements in ISO 9001 with surveillance audits, and in the third year a complete reaudit of the quality system.

Secondary

4. Monitoring supplier's management review activities.
5. Mutual recognition and acceptance of software registration among accredited registrars.
6. Use of software registration logo at option of supplier duly registered. Logo and name, if any, should in no way imply that this will apply in any manner to separate registrars, separate audits, separate fees.

The primary elements comprise the heart of the program and are essential to assure the program benefits are achieved. These have to do with assuring auditors are properly trained for software quality systems auditing, that they have the requisite workplace experience, and that they are used by registrars when the registration scope includes software development regardless of the industrial sector involved. The importance of the registration cycle is based on the fast pace of the information technology (IT) technology and change of organization, people, and processes, which make a 3-year cycle reasonable for a full account of varying conditions in individual supplier circumstances in terms of the reregistration.

The secondary elements are those that are important but not fundamental to the benefits of the program. For example, monitoring supplier's management review activities is an element, but experience has shown that not all registrars have always checked this item. This item is included to emphasize its importance, but it could be dropped from this particular program.

Mutual recognition and acceptance of software registrations among accredited registrars is included, since experience in the United Kingdom has shown that registrars often did not accept software registrations when they were called in to do a registration in a company that had used other registrars in the past. This practice resulted in reregistrations by the new registrars in areas where the original certificate(s) was still valid. The practice was actually a commercial abuse by registrars to get more revenue. Traditionally, mutual recognition and acceptance of certificates has always occurred through direct negotiations between certification bodies in the free market. This particular element could be further investigated and solved at the accreditation level by ANSI-RAB and dropped from the program if suppliers would not object.

Finally, use of the program name and logo (mark) in the SQSR program is still an open question. The *Guide* does not specify it. The implication here is that the matter is left to the market forces to make the ultimate decision.

Concerning National Systems and International Systems

One of the arguments against the SQSR program asserts that a national system is a barrier to trade and should not be allowed. On the other hand, the same argument recommends that work should be done at the international level, preferably in ISO/TC176.

These statements reveal a lack of understanding of the different scopes of operation of standards development bodies and conformity assessment bodies and their interactions at national, regional, and international levels. Different bodies are responsible for international standards and conformity assessment guides. Quality systems registration is the most recent branch of conformity assessments, and the necessary infrastructures at national and international levels are incomplete and are still being established. The critics also ignore the nascent and fluid nature of the relationships that exist between national and international systems. It is necessary to go back a bit into the historical evolution of conformity assessment internationally.

Product certification and laboratory testing have long been in place around the world in different countries, and these programs have existed at the level of the individual certification bodies and laboratories. Acceptance of certificates and test results was carried out via memoranda of understanding between the specific bodies themselves. Accreditation bodies have only recently come into existence. Systems were developed at the level of certification and laboratory testing first. National systems then began developing in a select number of countries relatively recently, but this is now taking place at a rapid pace, and only in the last several years has there been a focused movement to put in place international mutual recognition and harmonized practices using accreditation bodies as the pivotal bodies.

As of this writing, there is no applicable body at the international level with responsibility and authority to develop a program of this nature. ISO/TC176 writes the quality management standards, but it does not have the authority to develop application systems in which its standards are used for conformity assessment in accreditation and registration at either the international or national level. ISO/CASCO, coordinating with the IEC, has the charter to write guides in the conformity assessment field, but it does not have authority to implement particular accreditation or certification operating programs at either the national or international level. The absence of any existing international body to develop these systems to satisfy a pressing need has given birth to three new informal international activities: the International Accreditation

Forum (IAF), ISO itself, and ITQS. The United States is represented in all three efforts.

SOFTWARE QUALITY ASSURANCE

Computers and microprocessors are part of our world, and we are finding new applications of computers in products and in data processing. Such is the influence of the *computer age* that both at work and in the home we seem to depend on computers from the simplest to the most complex applications. The implication for those of us involved in quality assurance assessments is that an increasing number of design, manufacturing, and service organizations will be designing, installing, and operating software-based systems, and therefore software quality assurance will often need to be included within the scope of an assessment.

In addition, as hardware design and manufacture increasingly benefit from the effects of automation and new technologies in improving productivity and therefore reducing costs, software design remains a very labor-intensive activity with the result that the number of people employed in software engineering is continually increasing in proportion to those engaged in hardware design and manufacture. Also, the low cost and versatility of small computers is opening up many new applications for software-based systems at the expense of hardware solutions.

All this poses a problem for those people who belong to a generation brought up on hardware who are having to adapt to the software age. The problem is fundamentally one of communication. The fact that the world of software and computers is seen to have a language all its own, which is regarded as jargon by outsiders or nonspecialists, is the proof.

The Jargon should not present a problem between the assessor and the organization being assessed, because the assessment standards themselves are often common to those for hardware quality systems and therefore are written in familiar language. There are, however, some special terms that we should define.

Sector: As the term is used by the Software Committee, sectors are industries. Examples of sectors are the automotive industry and chemical industry. It is helpful to visualize an industry sector such as chemicals, electronics, or information technology (IT) as vertical columns in a matrix. Software is a discipline or specialty that may be applied in any industry. It can be visualized as a horizontal row in the same matrix since it exists in virtually all industries and therefore plays a role in all industry sectors. The U.S. Software Program is therefore not a sector program as the term is currently used by the RAB.

Program: The SQSR Committee uses the word *program* to describe the packaging of the expansion in all related aspects: auditors, registrars, and accreditation.

Arrangements: This word has been suggested as another term that could be used instead of program, but the word has already been used to describe the basic set of accreditation elements in the program.

Scope: This word has also been suggested as a term to replace sector, but the word *scope* already has two quite distinct usages and a third usage might add to confusion in the language. The first usage is that of the industrial fields a registrar is accredited to, such as electronics, plastics manufacturing, etc. The second usage is for the application or breadth of the supplier's quality system that the registrar audits and attests to on the registration certificate.

Software: Software covers all instructions and data that are input to a computer to cause it to function in any preset mode; it includes operating systems, compilers, and text routines as well as application programs. The definition includes the documents used to define and describe the program (including flowcharts, network diagrams, and program listings) and also covers any associated specifications, test plans, test data, test results, and user instructions.

Configuration: The complete description of the product and the interrelationship of its constituent elements. The configuration of a product is a collection of its descriptive and governing characteristics that can be expressed in functional and physical terms. The functional terms express the performance that the item is expected to achieve; the physical terms express the physical appearance and composition of the item.

Configuration Baseline: A specific reference point of product definition to which changes or proposed changes may be related.

Configuration Control: The discipline that ensures that any proposed change and addition (modification or addition) to the configuration baseline shall be prepared, accepted, and controlled in accordance with set procedures. This is sometimes called *modification control.*

Software Standards: Compared with the several thousand standards that are published nationally and internationally for hardware i.e., components, materials, very few standards have been published that define and describe the processes of software design and development. Likewise, very few companies engaged in software have either adopted published standards or written their methods. Therefore, many companies have become overreliant on individuals, so that when people leave the company it is difficult to understand their software designs. In these circumstances, software maintenance becomes prohibitively expensive.

Standards should be applied to achieve uniformity in the product so that others can interpret the design and therefore carry out improvements (enhancements) or error corrections (debugging).

Current standards include the following:

BS 3527	Glossary of terms
BS 4058	Flowcharts
BS 5476	Network diagrams
BS 5515	Documentation
BS 6719	Specification of user requirements
AQAP-13	Software quality system requirements
DEF STAN 00-16	Quality assurance of software

Software Life Cycle

Software life cycle (project life cycle) is defined as the totality of stages through which a software system passes. Typical phases are shown in Table 8.2.

As a general rule, the conception phase is a very small part of the overall life cycle. For this reason, very often perhaps a greater investment (of time, money, or other resources) for achieving a definitive, realistic (in every aspect) requirement would considerably reduce costs in the production and maintenance phases.

Table 8.2 Phases of the Software Life Cycle

Conception	Idea	Is the idea worthwhile?
	Specifications	Is it achievable?
	Feasibility	Is it effective?
	Requirement	Is the requirement definitive?
Production	Design	System design
	Development	Hierarchy of design modules
	Testing and approval	Integration
	Integration	Develop and test code
	Customer acceptance	Maintenance documentation
		User documentation
Maintenance	Operation	Debugging (error correction)
	Support	Enhancements
	Modification	Changes
	Enhancement	Change control (from/to any stage)
		Review
		Are objectives met?
		Is development under control?
		Is completion on target?
		Are standards/codes of practice being applied?

Hardware/Software Differences

When embodied in a product, software is analogous to hardware insofar as it is a component of the product. The first stage of assembly of software into the product is to produce firmware by entering the software into a storage device. The firmware is then assembled into a hardware assembly along with hardware components. Therefore, in the hierarchy of the total product, software via firmware may be considered a building block in the same way as hardware components.

There are, however, some differences between software and hardware of which the assessor should be aware because of their implementations on the required conduct of an assessment.

The major difference between software and hardware is that software is almost entirely a design and development activity. The term *replica-*

tion is used to define the reproduction of software, but it is a straightforward copying process and does not have the multitude of potential failure mechanisms inherent in hardware manufacturing.

Therefore, software quality assessment is concerned mostly with design controls and change controls, testing, and documentation. Also, software programs are normally organized on a project rather than a functional basis, with one team being responsible for all stages of a particular project.

OVERVIEW OF THE GUIDELINES

As already mentioned, it is important to recognize that the ISO 9000-3 is not a certifiable standard. Rather, it is a guideline for all software services to follow in their quality quest for continual improvement. The guideline provides the basic blocks of quality for all software services and it encourages them to focus on customer satisfaction.

The guidelines provide a systematic method of defining *what* software and hardware are and *how* the organization can go about improving the quality system in the short and long term.

All software organizations are included, if they are indeed interested in quality. All software organizations can improve, if they are indeed focused on the customer and its needs. The ISO structure provides the path for that improvement.

THIRD-PARTY REGISTRATION FOR THE SOFTWARE INDUSTRY

Third-party registration to ISO 9000 came to the United States in the late 1980s, and the RAB began to establish itself in 1989. The rationale for third-party registration is that it substitutes the objective, impartial, and competent service for first-party claims that are partial. Third-party surveillances can be used by the supplier as an external check, using the ISO 9000 quality system as the platform for continual quality improvement. It then allows the supplier to leverage the use of the registration in building the customer's confidence in a more cost-effective way over the entire marketing program.

This registration is especially useful in the global marketplace where buyers and suppliers may not physically meet. Buyers are able to purchase with more confidence that the supplier has a quality system and is consistently following it. The third-party system brings considerable savings to both customers and vendors by cutting down on the need for second-party audits by each buyer. To give greater confidence to all parties, accreditation bodies were formed to accredit registrars by thorough review and audit of the registrars to assure they are competent and conduct their business with suppliers according to rules that assure the system is operating impartially and competently for the benefit of the suppliers, purchasers, and public at large.

THE CASE FOR A COMPUTER SOFTWARE QUALITY SYSTEM REGISTRATION PROGRAM

Added Value

The software quality system registration program is intended to add value to the ANSI-RAB accreditation program as well as to the U.S. registration system currently operating. It would benefit both the suppliers and customers where computer software is the product or is included in the system delivered. It would not impose any additional requirements on the suppliers since ISO 9001 remains the registration standard.

The heart of the program is the requirement that auditors used in the process by registrars are properly qualified and meet a consistent standard of stated minimum qualifications and workplace experience both in software development and in quality auditing. Another important feature of the program is the use of ISO 9000-3, a guideline standard especially written to interpret ISO 9001 for software. This recognized guidance document puts the clauses of ISO 9001 in the language of software developers. The software program would contribute to acceptance of U.S. suppliers' registration of their quality systems where software is involved if it is deemed equivalent to those systems already operating in Europe.

Objective Qualifications

The present accreditation system used in the ANSI-RAB program is generic for all industries. The Software Quality System Registration (SQSR) program would establish objective qualifications for use in assessment of applicants in the auditor certification process. Minimum workplace experience in software development and in software quality is stated in the recommendations for the three grades of software auditor.

The SQSR recommendations give the present systems of ANSI-RAB registrar accreditation and RAB auditor certification extra credibility for the software program. They give an objective standard that could be used in both the accreditation process for registrars and the certification process for auditors. As a result, the overall system of registration would give the ANSI-RAB credentials increased credibility, which would in turn be beneficial to suppliers and customers. This would also contribute to enhanced reputation at the international level with direct benefits to software suppliers in this country whose registration carry the ANSI-RAB mark.

Registrars and Auditors

The specific program requirements are placed on the registrars and auditors, not on suppliers. This program enhances the requirements for accreditation of registrars who wish to expand their scopes of operation into the field of software development. The program would assure benefit to the suppliers and customers through standardized requirements for qualifications of the auditors. These requirements would give confidence to the suppliers and customers that results of the audits and registrations by different auditors and registrars would be equivalent, not only in the U.S. market but even more so in the international and global market, where buyers and suppliers may not meet. Results must be seen to be equivalent for the registration certificate to have any significance internationally.

Registration Is Voluntary

It is a matter of fact that the market forces have driven many suppliers to choose registration. This is not to say that every supplier must be registered. Except for regulated products where third-party registration or certification may play a role, it is the customers and marketplace that determine whether the supplier should be registered. Third-party registration is voluntary. If suppliers wish to be registered, they make that decision based on their customer requirements and market situation. Third-party registration should be regarded as a value-added service that is available for the market where suppliers and customers find it useful.

Some companies feel—and rightfully so—that they have progressed beyond ISO 9000 in their internal efforts to attain quality in their processes and products. Some believe that they should not be required to submit to ISO 9000 registration. As stated above, both the use of the standards themselves and registration are voluntary. The arguments over whether generic ISO 9000 registration should exist should not be allowed to color the situation with respect to software. Market forces in the global market have now settled the question of registration's existence. If some companies wish to remain outside its application, that is their decision. Many other companies, however, find it useful as a platform to continually improve internal quality and simultaneously demonstrate to the marketplace that they are doing so. The ISO 9000 quality standards are grounded on good business practice and quality principles that have evolved over many years in well-managed companies.

The issue is how to add value where registration is desired by the suppliers and customers, and where software is involved in the product. ISO 9001 does not mention software, and the language used is different from that used by software developers. Experience proves that without some added guidance and specific standards to be used with ISO 9001, there is risk that auditors will not be properly qualified and software developers will receive no added value in the audit or registration process. This was the experience that the United Kingdom had with third-party registration prior to their implementation of an enhanced program, which was subsequently called TickIT. Although TickIT has critics, it has many strong followers in both the United Kingdom and other countries, including the United States, for the added value it gives in the quality of the auditors and resulting audits. TickIT's basic intent cannot be faulted: to match software auditor experience to the customer needs (those of the supplier applying for registration) and add value to the supplier by performing audits that will raise the issues in areas requiring the supplier's additional attention in eliminating nonconformities as well as confirming that the quality system is operating effectively.

QUALITY MANAGEMENT AND QUALITY ASSURANCE STANDARDS

The aim of this section is to summarize each of the clauses in the guidelines and to provide the reader with an overview of what software quality is all about from the ISO perspective.

Section 1:   Scope.   This guideline establishes the need for implementing a quality system in the software industry. Specifically, this guideline sets out to facilitate the application of ISO 9001 to organizations developing, supplying, and maintaining software.

Section 2:   Normative References.   This guideline provides for the references that are used to define the service quality.

Section 3:   Definitions.   In this section the specific vocabulary for the software is utilized and is based in the ISO 2382-1 and ISO 8402.

Section 4:   Quality System—Framework.   In this section the guideline provides for:

Section 4.1:   Management Responsibility

**Section 4.1.1:   Supplier's Management
                       Responsibility**

Section 4.1.1.1:   Quality Policy.   This section requires that the supplier's management shall define and document the objectives and commitment to quality at all levels of the organization.

Section 4.1.1.2:   Organization

Section 4.1.1.2.1:   Responsibility and Authority.   This section requires that the authority and responsibility for those who perform or are involved—on any level—with quality should be defined. Some specific guidelines—to help identify the personnel responsible—are given for direction. They are:

- Personnel who initiate action for prevention purposes
- Personnel who identify or record any quality problems
- Personnel who have the power or position to initiate, recommend, and provide solutions
- Personnel who verify implemented solutions
- Personnel who are authorized to control the nonconformities in either processing, delivery, or installation until such a time when the nonconformities have been corrected

Section 4.1.1.2.2: Verification Resources and Personnel. This section provides for in-house verification requirements. The identification of the requirements will depend on the supplier. Examples for some of the criteria are given in positions like:

- Inspection
- Test
- Production
- Installation
- Servicing

One of the most important aspects of this section is the second paragraph, which states that the verification requirements *shall be carried out by personnel independent of those having direct responsibility for the work being performed.*

The emphasis of this section is indeed on an independent verification, which, of course, is the thrust of the ISO 9001 standard in clause 4.1.2.2.

Section 4.1.1.2.3: Management Representative. This clause reiterates ISO 9001 clause 4.1.2.3, which states that the supplier must appoint a management representative who has the authority and responsibility for ensuring the quality system is implemented and maintained.

Section 4.1.1.3: Management Review. This section calls for management to define an appropriate interval for reviewing the suitability and effectiveness of the requirements set by the ISO 9001 standard. In addition, this section reemphasizes the need for records for such activities.

Section 4.1.2: Purchaser's Management Responsibility. This section calls for cooperation between purchaser and supplier to provide the necessary information for all pending concerns. It also gives some guidelines, including:

- The definition of the requirements
- A system for answering questions from the supplier
- A system for concluding agreements with the supplier
- A system that ensures the agreement(s)
- A system that defines the acceptance criteria
- A system that deals with software items unsuitable for use

Section 4.1.3: Joint Reviews. This section requires that the supplier and purchaser shall have documented "regular joint reviews" to cover the following:

- Conformance to the requirements
- Verification of results
- Acceptance of results

Section 4.2: Quality System

Section 4.2.1: General. This section requires that a quality system must be established, maintained, and documented. The focus of this quality system must be prevention, and as such it must be instituted throughout the life cycle.

Section 4.2.2: Quality System Documentation. This section provides for all quality system elements to be clearly documented in an orderly manner.

Section 4.2.3: Quality Plan. This section provides for all quality activities to be documented in a quality plan. This plan should be understood and followed by all concerned.

Section 4.3: Internal Quality System Audits. This section echoes the requirements of ISO 9001 clause 4.17 for a comprehensive planned and documented internal quality audit to verify the compliance and effectiveness of the quality system. It is important to note that just like ISO 9001 clause 4.17, this section of the guideline requires that the results be reviewed by management and appropriately documented and, where applicable, the appropriate corrective action should be taken.

Section 4.4: Corrective Action. This section establishes the need for corrective action and provides specific steps for such action. They are:

- Investigate the cause and corrective action of a problem
- Analyze all applicable information to identify and remove potential problems
- Initiate preventive actions to deal appropriately with the problem
- Apply the appropriate controls to ensure that corrective measures are taken and that these measures are effective
- Implement and record all changes in the appropriate procedures as a result of the specific corrective action and notify all appropriate personnel of the changes

Section 5: Quality System—Life Cycle Activities

Section 5.1: General. The recommendation that the life cycle model should be followed in the software development is the essence of this clause.

Section 5.2: Contract Review

Section 5.2.1: General. According to this clause, it is the responsibility of the supplier to establish and maintain procedures for contract review. Each contract should be reviewed to ensure that:

- The scope and requirements are defined and documented
- Contingencies or risks are identified
- Proprietary information is protected
- Differences are resolved
- Supplier capability is assured

- Subsupplier capability is accounted for
- Terminology is understood
- Contractual obligations can be met

Records should be maintained for all the above.

Section 5.2.2:   Contract Items on Quality.   This clause identifies specific items frequently found in a contract.

- Acceptance criteria
- Handling of charges
- Handling of problems
- Activities carried out by the purchaser
- Facilities, tools, and software items
- Standards and procedures to be used
- Replication requirements

Section 5.3:   Purchaser's Requirements Specification

Section 5.3.1:   General.   This clause mandates that with software development, the supplier should have a complete, unambiguous set of functional requirements as well as appropriate records.

Section 5.3.2:   Mutual Cooperation.   This clause provides recommendations to be followed by the development of the purchaser's requirements specification.

- Assignment of persons
- Methods for agreeing
- Efforts to prevent misunderstandings
- Appropriate documentation

Section 5.4:   Developing Planning

Section 5.4.1:   General.   This clause develops a plan based on the following:

- The definition of the project
- The organization of the project resources

* Development phases
* The project scheduling
* Identification of related plans

Section 5.4.2   Development

Section 5.4.2.1:   Phases.   To optimize the development plan, this clause proposes a division of work into phases and the identification of:

* Development phases
* Required inputs
* Required outputs
* Verification procedures
* Analysis of the potential problems

Section 5.4.2.2:   Management.   The development plan should define how the project is to be managed, including:

* Schedule and development
* Progress control
* Organizational responsibilities
* Organizational and technical interfaces between different groups

Section 5.4.2.3:   Development Methods and Tools.   According to this clause, specific methods for ensuring that all activities are carried out correctly must be developed. They may include:

* Rules, practices, and conventions
* Tools and techniques
* Configuration management

Section 5.4.3:   Progress Control.   All progress reviews should be planned.

Section 5.4.4:   Input to Development Phases.   For each development phase the required input should be defined and documented.

Section 5.4.5: Output from Development Phases. This clause requires output from each development phase that should be defined and documented. The output should be verified and should:

- Meet the requirements
- Contain or reference acceptance criteria
- Conform to appropriate practices and conventions
- Identify those characteristics of the product that are crucial to its safe and proper functioning
- Conform to applicable regulatory requirements

Section 5.4.6: Verification of Each Phase. This clause recommends that the supplier should draw up a plan for verification of all development phase outputs at the end of each phase. Control measures should be based on:

- Holding development reviews
- Comparing designs
- Undertaking tests and demonstrations

Section 5.5: Quality Planning

Section 5.5.1: General. This clause emphasizes the need of a quality plan by the supplier.

Section 5.5.2: Quality Plan Content. This clause defines the requirements for the quality plan as:

- Quality objectives
- Defined input and output criteria
- Identification of types of test
- Detailed planning of test
- Any specific responsibilities

Section 5.6: Design and Implementation

Section 5.6.1: General. This clause defines the design and implementation activities that transform the purchaser's requirements specification into a software product.

Section 5.6.2: Design. This clause defines aspects inherent to the design activities that should be taken into account, such as:

- Identification of design characteristics
- Design methodology
- Use of historical data
- Subsequent processes

Section 5.6.3: Implementation. This clause defines additional requirements common to all development activities, such as:

- Rules
- Implementation methodologies

Section 5.6.4: Reviews. This clause requires that the supplier should carry out reviews to ensure that the requirements are met and the above methods are correctly carried out.

Section 5.7: Testing and Validation

Section 5.7.1: General. This clause defines the need for testing and integration. However, the specific approach is left to the individual organization.

Section 5.7.2: Test Planning. This clause requires the establishment and review of the testing plans, specifications, and procedures before the testing activities are begun. In addition, it provides some guidelines. They are:

- The plans for the software item, integration, system and acceptance test
- Test cases, data, and expected results
- Types of tests to be performed
- Test environment
- The criteria for the test
- User documentation
- Personnel requirements

Section 5.7.3: Testing. This clause recommends some special considerations within testing. They are:

- The test results should be in tune with the requirements
- Any problem and its possible impact should be recorded and the appropriate personnel should be notified. All problems should be tracked until they are solved.
- Retesting should be done whenever modifications have occurred
- Test adequacy and relevancy should be evaluated
- Consideration and documentation should be given to the hardware and software

Section 5.7.4: Validation. This section emphasizes the need and place for validation. Specifically, it calls for validation before delivery of the product and under conditions similar to the application environment.

Section 5.7.5: Field Testing. This section recommends that if field testing is necessary, then:

- The individual features should be tested in the field environment
- The specific requirements between supplier and purchaser should be identified
- (After test) responsibility of restoring the user environment is required

Section 5.8: Acceptance

Section 5.8.1: General. This section provides some safeguards at the time of delivery. They are:

- Delivery should be accepted based on previously agreed criteria
- The method of handling problems (and their disposition) during acceptance should be agreed on before the delivery and it should be documented

Section 5.8.2: Acceptance Test Planning. This section provides some guidelines for section 5.8.1. They are:

- Time schedule
- Methods of evaluation
- Environment and resources
- Acceptance criteria

Section 5.9: Replication, Delivery, and Installation

Section 5.9.1: Replication. This section recommends several considerations prior to delivery. They are:

- Number of copies
- Type of media
- Manuals and user guides
- Copyright and or license
- Custody of master and/or backup copy
- The obligation (how, what) of the supplier to the purchaser

Section 5.9.2: Delivery. This section reiterates the need for acceptance criteria.

Section 5.9.3: Installation. This section provides some guidelines for the installation process. They are:

- Schedule
- Access to purchaser's facilities
- Skilled personnel
- Access to purchaser's system and equipment
- Validation consideration
- Formal procedures for installation

Section 5.10: Maintenance

Section 5.10.1: General. This section provides typical maintenance considerations after delivery. They are:

- Problem resolution
- Interface modification
- Expansion or modification or improvement of the system

- If specific items are under consideration, then they should be identified on a contractual agreement. Examples are:
    Item identification
    Period of time
    Program
    Data and their structures
    Specifications
    Documents for purchaser and/or user
    Documents for supplier's use

Section 5.10.2:   Maintenance Plan.   This section provides some guidelines on maintenance activities. It reiterates the need for contractual agreement before delivery and it establishes some minimum criteria for the plan. They are:

- Scope of maintenance
- Identification of the product
- Support needed
- Maintenance activities
- Maintenance records and reports

Section 5.10.3:   Identification of the Initial Status of the Product. This section defines the need for the identification as well as the need for having it documented and agreed upon by all parties concerned.

Section 5.10.4:   Support Organization.   This section recommends the establishment of an organizational structure between suppliers and purchasers to support the maintenance activities. If the structure is established, it must be flexible enough to address the unexpected occurrence of problems.

Section 5.10.5:   Types of Maintenance Activities.   This section emphasizes the need for consistency throughout all the changes in the software. These changes must not only be based on current documentation, but they must also be in accordance with document control and configuration management. The flow of the standardization may be done through at least the following consistent activities:

- Problem resolution

- Interface modification
- Functional expansion or performance improvement

Section 5.10.6: Maintenance Records and Reports. This section defines the need for all maintenance activities to be recorded in standardized formats and retained. Some of the information that these records should contain includes:

- List of requests
- Problem reports
- Personnel responsible for identifying the problem and implementing the corrective action
- Priorities to the corrective actions
- Results of the corrective actions
- Statistical data for failures and maintenance activities

Section 5.10.7: Release Procedures. This section recommends some procedures for incorporating changes in a software product. They are:

- Set ground rules for all pending releases
- Description of the types or classes of releases
- Methods for future changes
- Methods to confirm that changes implemented will not create other problems
- (For multiple products and sites) requirements for records indicating which changes (where and what) have been implemented.

Section 6:   Quality System—Supporting Activities (Not Phase-Dependent)

Section 6.1   Configuration Management

Section 6.1.1:   General. This section defines configuration management as a mechanism for *identifying, controlling, and tracking the versions of each software item*. In addition, it gives some of the fundamental characteristics that configuration management should include. They are:

- Identify uniquely the versions of each item
- Identify uniquely the versions of a complete product

- Identify the build status of software products
- Control simultaneous updating of a given product
- Provide coordination for the updating of multiple products
- Identify and track all actions and changes

Section 6.1.2: Configuration Management Plan. This section recommends a configuration plan with the following:

- Responsibilities assigned
- Activities to be carried out
- Tools, techniques, and methodologies to be used
- The stage at which configuration management takes over

Section 6.1.3: Configuration Management Activities

Section 6.1.3.1: Configuration Identification and Traceability. This section recommends that appropriate identification should be established and maintained during all phases of software configuration. Typical points of identification and traceability are:

- Functional and technical specifications
- All development tools
- All interfaces
- All documents and computer files

Section 6.1.3.2: Change Control. This section reiterates the need for appropriate document control in establishing and maintaining appropriate procedures under configuration management.

Section 6.1.3.3: Configuration Status Report. This section recommends that appropriate records should be established and maintained to record, manage, and report the status of software items.

Section 6.2: Document Control

Section 6.2.1: General. This section supplements the requirements of ISO 9001 under document control. The issue in this section is the appropriateness of establishing and maintaining the documentation. Specifically, this section covers:

- The determination of "appropriate" documents
- The approval and issuing procedures
- The change procedure

Section 6.2.2: Types of Documents. This section categorizes some of the documents into the following:

- Procedural documents
- Planning documents
- Product documents

Section 6.2.3: Document Approval and Issues. This section recommends that all documents should be reviewed and approved by authorized personnel prior to issue. Specifically, procedures should exist to ensure that:

- All appropriate documents are available at appropriate locations
- All obsolete documents are properly removed

Section 6.2.4: Document Changes. This section echoes the requirements of ISO 9001 clause 4.5.2. Specifically, it emphasizes the need that all changes must be reviewed and approved and that a master list or equivalent document control must be in place to identify the current version of documents in order to preclude the use of nonapplicable documents.

Section 6.3: Quality Records. This section echoes again the requirements of ISO 9001 clause 4.16. Specifically, it emphasizes that the need for establishing and maintaining procedures for identification, collection, indexing, filling, storage, maintenance, and disposition of all quality records is the supplier's responsibility.

The quality records should be legible and identifiable to the product involved. They should also be current.

Section 6.4: Measurement

Section 6.4.1: Product Measurement. This section of the guideline recommends that at a minimum, some metrics should be used to quantify the software quality. (It is very important to realize that the guideline recognizes that there are no universally accepted measures for software quality.) In addition, this section recommends that the quantifiable measures should be used for the following purposes:

- To collect data and report the quantification
- To identify the current level of performance
- To take remedial action when necessary
- To establish specific improvement goals

Section 6.4.2:   Process Measurement.   This section addresses the issue of quality and measurability of the development and delivery process. It proposes measurement of:

- How well the planning and development is being monitored
- How well the planning and development is being carried out
- How well the planning and development is being met on schedule
- How effective the development process is at reducing the probability that faults are introduced.

In addition, this section recommends that process control be used, however, it should fit the appropriateness rule.

Section 6.5:   Rules, Practices and Conventions.   This section encourages the supplier to have rules, practices, and conventions that support the ISO 9000 standard. Furthermore, these rules, practices, and conventions should be reviewed and revised as required.

Section 6.6:   Tools and Techniques.   This section encourages the supplier to use appropriate tools, facilities, and techniques so that they support effectively the ISO 9000 standard. The effectiveness may be for management purposes as well as product development. Improvements should be made as required.

Section 6.7:   Purchasing

Section 6.7.1:   General.   This section recommends that the supplier should ensure that a purchased product or service conforms to specified requirements. The documents should contain data clearly describing the product or service ordered.

Section 6.7.2: Assessment of Subcontractors. This section echoes ISO 9001 clause 4.6.2. Specifically, the subcontractors should be selected on the basis of their ability to meet the requirements, including quality requirements.

Section 6.7.3: Validation of Purchased Product. This section reaffirms that the supplier is responsible for the validation of subcontracted work. Of course, all special and/or specific requirements should be contractually agreed upon.

Section 6.8: Included Software Product. This section recommends specific action from the supplier with regard to quality requirements. They are:

- Procedures for establishing and maintaining:
  Validation
  Storage
  Protection
  Maintenance

For all the above, consideration should be giving to the support of the software product.

Section 6.9: Training. This section recommends establishing and maintaining procedures for identifying the training needs and providing for the training of all personnel performing activities affecting quality. Specifically, this section suggests that the personnel with specific tasks should be qualified and the appropriate documentation should exist and be maintained for such specific tasks.

IMPLEMENTATION OF THE ISO STANDARDS IN THE SOFTWARE INDUSTRY

The implementation process for the ISO standards in any software industry is exactly the same as in any other organization. The steps of implementation are the same as those identified in Chapter 5.

AUDIT IN THE SOFTWARE ORGANIZATION
TO THE ISO STANDARDS

As of this writing, there is no harmonized international system for audit. As a consequence, the audits for the software industry can follow two paths: (a) the TickIT—a de facto international system, or (b) the traditional ISO 9001 approach.

What is special about the software audit is that one must understand that while the hardware assessments tend to have horizontal or departmental audits, the software assessment will focus more on a vertical audit, which will be more appropriate. By doing a vertical audit the assessor assures that the various phases of a project can be assessed with particular regard to the software life cycle.

CURRENT ISSUES IN THE SOFTWARE
INDUSTRY

There are at least three major issues in the software industry as they relate to certification: (1) supply and demand, (2) bureaucracy, and (3) RAB's responsibility.

Supply and Demand

One of the arguments leveled against the SQSR program is that it will add costs, complexity, additional requirements, and bureaucracy to ISO 9000 registration for suppliers and these added costs will be passed on to customers. This argument is faulty on several grounds.

The cost component of the argument is based on the fact that TickIT audits in the United States are more expensive than generic registrations. It is true that TickIT audits are more expensive because all the auditors must be certified by the IQA/RBA since the United States has no equivalent program and we cannot generate the required number of auditors in this country.

(There are Americans who have successfully taken the TickIT course and who have applied to RAB for certification. RAB has thus far returned their applications with the statement that the applicants do not qualify for certification on the grounds that RAB does not recognize the TickIT training courses as being equivalent to RAB-recognized ISO lead assessor courses. In the present circumstance, the U.S. system cannot generate auditors who can be used even by the U.K. registrars in TickIT audits in this country. Some Americans have applied to IQA for certification as TickIT auditors, but this is a convoluted and costly way to make auditors available in our own market.)

The final point to be noted here is that there is considerable consensus both in the United States and internationally among well-informed observers that TickIT and other IT-specific programs will not just *go away*. U.S. IT suppliers and their customers will become increasingly aware of the benefits of such a program. As they attempt to satisfy their desire to have this type of registration, the opportunities without an equivalent U.S. program will be very limited because there will be only a few U.K. registrars in the United States to perform the service. It stands to reason that the U.S. suppliers will pay more at that time and wait longer for such registrations—and thus be at a competitive disadvantage in the United States and world markets.

Bureaucracy

The responsibility of bureaucracy is with the RAB. However, there are at least two questions—regarding additional costs and bureaucracy and second level of registration for suppliers—that persist from the critics of both RAB and SQSR.

RAB's Responsibility

While the objective of the ANSI-RAB program is to put our country on a level playing field with the rest of the world with respect to ISO 9000 quality systems accreditation of registrars, the RAB is responsible for approving training course providers and certification of ISO 9000 auditors.

The ANSI and RAB have a responsibility to act in the interests of all legitimate parties. After all, it is a fact that many other countries are carefully studying what is going on in this field and are preparing recommendations for implementation. It is a fact that TickIT and ITQS players are seriously engaged in a harmonization process. It is a fact that the United States risks being left behind. It is a fact that segments of the

IT industry that want an equivalent system and a mechanism for playing a role at the international table are being denied that opportunity by the absence of an equivalent system in the United States.

As of this writing, it appears, however, that there is no harmonization agreement between all the concerned parties. It is hoped, however, that both ANSI and RAB along with software producers will recognize there is a need for the recommended program in the overall interest of the United States even though support for it may not be unanimous.

SPECIAL CONSIDERATIONS IN TRAINING AND CERTIFICATION*

Because the RAB has decided to proceed with implementation of a software sector program, let us look at the ramifications of such an activity. Bear in mind that these may change, as this area is very volatile and discussions are still being held as of this writing.

In the international markets the focus for the software industry, at least for the certification and training part, seems to be the TickIT program. In the United States, however, as of this writing there has been no firm commitment to TickIT. Rather, there is an ongoing discussion about adopting the SQSR guide.

The essence of the TickIT guide is that it contains some information that is essential for it to be an accredited registration program. All the other information is simply there to help explain the program to purchasers, suppliers, and other general readers. The two sections that are there for advisory purposes are the purchaser and supplier sections. The introduction contains advisory and historical information, but it also includes the uniform accreditation arrangements, which contain essential elements of the program.

The TickIT uniform accreditation arrangements are defined as:

- Use of the TickIT name and logo by registrars and suppliers
- The registration cycle—3 years
- Monitoring the supplier's management review activities

---

*The information in this section is based on committee correspondence of the ASQC software division on SQSR from July 6, 1993 to March 25, 1994 and the author.

- Mutual recognition of accredited TickIT registration
- Use of TickIT guidance documentation
  ISO 9000-3
  European IT Auditors' Guide
  Auditor qualifications and certification requirements (using 10011 standards)
- Use of TickIT auditors by accredited registrars in registrations where software design and development are involved

In addition, it is implied that all accredited registrars offering this program will operate to EN 45012 (the general criteria for Registration Bodies Operating Quality Systems Registration). The registrar will audit to ISO 9001 using ISO 9000-3 as guidance on the interpretation of ISO 9001 for software. Auditor qualifications, certification, and registration auditing practices will be based on ISO 10011 standards.

The U.S. program has identical requirements for the uniform accreditation arrangements with those of TickIT. This is in order to assure equivalency of results and acceptance of certificates between the international community and the United States.

As of this writing, the difference between the TickIT and SQSR seems to be administrative, procedural, and political. However, the similarity is that both guides provide a mechanism for improving quality of software.

While the SQSR is still in the development stage, the TickIT is well established and provides the software industry with some guidelines for training. The training is usually a 5-day course covering the following items:

Day 1: Overview of quality
        The role of the auditor
        Overview of quality management system standards
        Documentation
        Workshops on ISO 9001 standards and quality problem identification
Day 2: Introduction of the concept that quality can be managed
        Software quality system
        Introduction of the TickIT guide

The audit system
The relationship between ISO 9001 and ISO 9000-3
Audit planning and preparation
Construction of checklists
Workshops on ISO 9000-3 guideline and preparing a check-
list
Day 3:   Audit responsibilities
Opening meeting
Investigation procedures
Reporting noncompliances
Review validation and verification procedures for software
Configuration management
Security and archiving
Workshops on practicing audit responsibilities and writing
noncompliances
Day 4:   Evaluating effectiveness
The closing meeting
Audit reporting
Corrective action and follow-up
How to handle revisions
Workshops on handling corrective actions and follow-up
Day 5:   Internal audit
Third-party assessments and surveillance requirements
Accreditation and TickIT requirements
Examination

REFERENCES

ISO 2382-1:1984. *Data Processing—Vocabulary*. Part 01: *Fundamental terms*.
ISO 8402:1986. *Quality—Vocabulary*.
ISO 8402: 1994. *Quality—Vocabulary*.
ISO 9000-3:1991. *Quality Management and Quality Assurance Standards: Guidelines for the Application of ISO 9001 to the Development, Supply and Maintenance of Software*.
ISO 9001:1994. *Quality Systems—Model for Quality Assurance in Design/ Development, Production, Installation, and Servicing*.
ISO 10011-1:1990. *Guidelines for Auditing Quality Systems*. Part 1: *Auditing*.

9
ISO and BS 7750

This chapter addresses the issues of the ISO quality standards and the environmental concerns as defined in BS 7750 in 1992. Our aim in summarizing the BS 7750 is to show that indeed *all* environmental issues can and should follow the building blocks of quality as identified by the ISO structure. A comparison between ISO 9001 and BS 7750 is shown in Table 9.1.

ENVIRONMENTAL QUALITY ASSURANCE

Perhaps the ultimate challenge facing world business now and in the years to come is the need to balance financial, quality, environmental, and legal responsibilities. Even though all these issues are very important, in this section we are concerned with the environmental issues.

As we begin to recognize the true cost of pollution, executives all over the world are realizing that environmental decisions can directly affect a company's financial future and may even affect survival.

Table 9.1 Comparison of ISO 9001 and BS 7750

Clause in ISO 9001	Description	Clause in BS 7750
4.1	Management responsibility	4.1, 4.2, 4.3, 4.11
4.2	Quality system	4.1, 4.7
4.3	Contract review	4.4, 4.5, 4.6
4.4	Design control	4.6, 4.7, 4.8
4.5	Document control	4.7
4.6	Purchasing	4.4, 4.8
4.7	Purchaser supplied product	4.4
4.8	Product identification and traceability	4.9
4.9	Process control	4.8
4.10	Inspection and testing	4.8
4.11	Inspection, measuring, and test equipment	4.8
4.12	Inspection and test status	4.8
4.13	Control of nonconforming product	4.8
4.14	Corrective action	4.8
4.15	Handling, storage, packaging preservation, and delivery	4.4, 4.8
4.16	Quality records	4.9
4.17	Internal quality audits	4.10
4.18	Training	4.3
4.19	Servicing	4.19, 4.8
4.20	Statistical techniques	4.8

Every country has some form of environmental laws and means of enforcing them. However, both the implementation and enforcement of these laws are not consistent throughout. The spectrum of variation is from Europe's strict laws, i.e., Germany's recycled and packaging directives, to Latin America's lax policies, i.e., Mexico's (Maquiladoras region) water pollution and health and safety policies, to almost non-existent laws, i.e., Brazil's policies on the rain forest.

If environmental impact assessment is important, then this inconsistency presents a problem, primarily because companies operating in countries with low regard for the environment are able to compete in the international markets on a favorable cost differential for their products.

To remedy the diversity of the international laws and their enforcement, the British Standards Institution (BSI) on April 6, 1992, launched the BS 7750 Environmental Management Systems, the first standard on the subject to be produced anywhere in the world. The intent of the system was to be, and remains, a sign of model management system for all types of organizations wishing to take a systematic and integrated approach to environmental performance.

It may be useful to point out the difference between standards and regulations (laws) and to relate national and EU standards to international standards. The EU publishes directives that member states may make compulsory with legislation such as ministerial regulations or acts of parliament. The health and safety directives issued by the EU have become compulsory regulations within member states. Somewhat confusingly, the EU may also publish regulations binding on member states, for example, environmental regulatory and accreditation schemes.

On the other hand, a national standard, such as BS 7750, which becomes an internationally accepted standard, or one already so accepted, such as ISO 9000, is a mechanism for meeting specified levels of performance and whole sets of regulations, and also of demonstrating conformance to both.

Through ISO, national standards become harmonized and made *standard* worldwide. ISO 9000 is the supreme example of a single universally (as of this writing, 91 countries) accepted quality management standard, for a system as distinct from a product or process (Rothery, 1993).

The aim of the BS 7750 standard is to accomplish the following initiatives within an organization:

1. Understand the benefits of environmental compliance: Leaders must know why corporate environmentalism is important to their industry, their company, their customer base, and society at large.
2. Set environmental goals: By defining specific environmental goals, progress can be measured and tracked. Environmental improvement can be accomplished if there is definition and measurement of goals.
3. Be environmentally responsible: Top management must be per-

sonally committed and they must make sure that all employees in the organization are environmentally accountable. Top management must also encourage the evaluation of outsiders in terms of environmental results. This means that it is management's responsibility to find out how the company performs from the customer's perspective, the government's perspective, and the public's perspective.

4. Earn the right to exist: Be both ecologically and fiscally responsible. Emphasize throughout the organization that *financial gain at the expense of the environment is totally unacceptable.*

5. Do the right thing: Top management must believe that their decisions will be supported by future stakeholders and the fundamental direction of environmental concerns will not change.

Remember, all environmental decisions are difficult. If the answer comes too easily, the wrong choice was probably made. Contrary to public opinion, the BS 7750 is not revolutionary in the sense of demanding a *new* system in your business or a *new* reorganization or *more* employees to document your environmental policies. Rather, the BS 7750 is a system that *adopts* (author's emphasis) a quality system approach to environmental issues. It is strongly linked to *existing* (author's emphasis) quality management practice as embodied in the series of standards known as ISO 9000. One, in fact may go as far as to say that the BS 7750 is a supplemental standard to the ISO 9000 series dealing specifically with environmental concerns. After all, the ISO 9000 deals with the processes that support quality, and BS 7750 deals with the processes that support the elimination of damage to the environment.

One of the questions often being asked about the standard is: "How does the BS 7750 standard define environment?" The standard defines environment as including human and other living systems within the surroundings and conditions in which they operate. It also encourages that the environmental policy be consistent with the occupational health and safety policy. The definition is scary; however, once you study it, it not only makes sense but it is quite appropriate. It emphasizes the TOTAL ecosystem.

What are some of the specific issues that the standard is really addressing? The following list provides some of the elements:

Air emissions	Water resources
Water supplies	Sewage treatment
Waste	Nuisances
Noise	Radiation
Urban renewal	Amenity, trees, and wildlife
Physical planning	Environmental impact assessment
Product use	Materials
Energy	Public safety
Staff health and safety	Chemicals

THE ESSENCE OF THE BS 7750

The standard, in the simplest form, is a standard to improve the environmental performance of an organization. It does this by forcing the organization to understand (as appropriate) the rules and demands of the environment and the community and then presents a method to satisfy them, by providing an outline for management to follow.

INSTALLING THE STANDARD

There are at least three reasons for a company to decide to apply for BS 7750 certification: government regulations, market demand, and marketing advantage. Of the three, the author believes that the marketing advantage is perhaps the most important for the near future. However, as time passes, government regulations and market demand will become more dominant advantages.

To install the standard with an ISO 9000 base, the organization has to add only the additional environmental clauses to the quality system. For a model of environmental implementation, see Figure 9.1. Note that quite a few of the ISO 9001 clauses are similar to the BS 7750 clauses. For a discussion of the detailed process of implementation, see Chapter 5.

To install the standard without an ISO 9000 base, the organization has to follow the normal implementation as explained in Chapter 5. In addition, the following steps must be taken.

1. Inventory ALL relevant environmental directives, regulations, and rules that affect the organization.

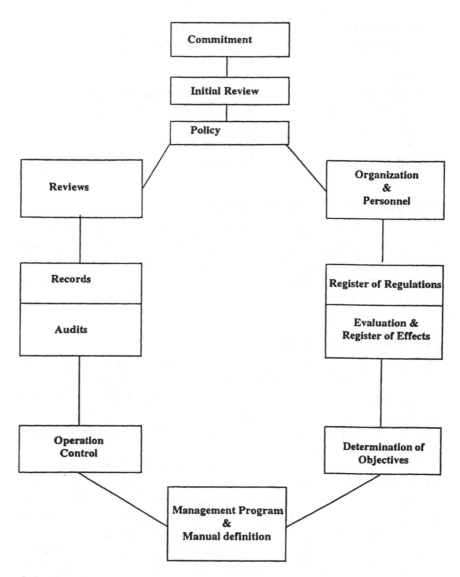

9.1 A typical sequential implementation process for the BS 7750. (Adapted from BS 7750.)

2. Develop the environmental management manual to include the following:

- Environmental policy
- Processes relevant to the standard
- Organization
- Cross-reference to the directives, regulations, and rules
- ISO 9000 relevant material
- Areas that are affected by the standard

3. Develop full documentation:

- All procedures identified
- All controls specified
- Design considerations
- Supplier assessment
- Design of audits
- Housekeeping
- Changes
- Inspection and testing
- All instructions identified
- Control documentation
- Forms

4. Appropriate and applicable training

COST AND BENEFIT

The subject of costs in the BS 7750 is uniquely different from that of traditional quality costs, where all costs involved apply to the costs of the quality control system and to the real costs of poor work, rework, and scrap. In the case of environmental costs, there is a cost both to the organization and to the environment itself.

Costs are mentioned in the standard in the context of providing sufficient resources, as a possible guide in setting environmental objectives and targets, and as a yardstick in the preparatory environmental review.

Surprisingly, Taguchi's theory of quality (the "loss function") suddenly comes into play with the new environmental management standard. Taguchi is the one who said that quality is measured as the loss society experiences after a product is shipped. Another way of looking at this is to think of maximum quality representing minimum loss. Certainly this notion can and does apply to the environmental question. Minimizing environmental impact and the wise use of resources equals both good environmental management and good quality, according to Taguchi's theory (Taguchi, 1987).

So, just like ISO 9000 and BS 7750, the world of quality assurance and environmental control are brought together.

Specifically, the project costs will be those incurred in designing and implementing the environmental management system and applying for certification.

The operations costs will be the annual costs of operating the environmental management system, including those of audits, upgrades, and annual recertification fees. Other running costs may be related to the costs of controlling, neutralizing, recycling, disposing, or otherwise treating the outputs of processes, including the costs of hardware and software.

On the other hand, in implementing the BS 7750 there are some potential benefits as well. They are:

Direct benefits:
    Reduction in resource consumption
        Energy
        Raw materials
    Reduction in waste production
    Reduction in employee and community complaints
    Reduction of industrial accidents
    Reduction of negligence claims
    Reduction of fines and penalties
    Reduction of personal liability
Indirect benefits:
    Enhanced corporate image
    Enhanced marketing capabilities
    Improved staff morale
    Improved customer relations
    Improved community relations

AUDIT TO THE BS 7750 STANDARDS

Since the audit is a systematic method of verification of the quality system, the approach to the BS 7750 environmental standards for the organizations that seek certification is the same as the one for ISO 9001.

In addition to the verifiable items that are part of the ISO 9001, the auditor must investigate the specific requirements included in the BS 7750. Examples, of those special items include:

Environmental policy
Conformance to governmental regulations
Policy for removal of hazardous waste
Policy for issues on pollution

THE FUTURE

As of this writing, the future of BS 7750 is uncertain. However, if history is any indication the ground is quite fertile for its adaption and its spread in the world markets. If the standard is the success that many expect it will be, it may replace the current ISO 9000 altogether, or ISO 9000 may simply expand to embrace it.

CHRONOLOGY OF DEVELOPMENTS IN THE ENVIRONMENTAL WORLD

ISO requires consensus to establish an international standard. Therefore, ISO/TC 207* ultimately will contain requirements with which all participating countries are in agreement. As of this writing, the following events have occurred (in reverse order).

---

*The ISO/TC 202 is organized into six subcommittees and one work group. ISO/TC 207 and each of its seven components are the designated responsibility of an ISO member. Subcommittees, in turn, are broken down into work groups, which also have designated national oversight. The organizational components are: SC1, environmental management system; SC2, environmental auditing and related environmental investigations; SC3, environmental labeling; SC4, environmental performance evaluation; SC5, life cycle assessment; SC6, terms and definitions; WG, environmental aspects in product standards.

- Present. The BS 7750 is in good position to be adopted as an environmental standard. If not as is, certainly either modified or expanded.
- May 1994. The ISO/TC 207 met in Australia for balloting. The resolution reached was to further study the issues and not make a binding decision yet. As of this writing, the committee has not announced its next meeting date and/or agenda.
- March 1994. The SCI met in Toronto, Canada, to discuss the elements for a new environmental management system proposed during the October 1993 meeting. No resolution was reached.
- October 1993. The SCI agreed to begin all over again with a clean slate. It agreed in favor of developing its own list of environmental management system elements and requirements. The elements were identified as:

    I. Policy
    II. Organization
    III. Legal and other requirements
    IV. Environmental effects
    V. Objectives and targets
    VI. Environmental management program
    VII. Operational control
        General
        New activities
        Emergency planning
        Noncompliance and corrective action
        Monitoring and testing
    VIII. Audits
    IX. Management review
    X. Documentation and records
        General
        Environmental manual
    XI. Training and motivation
    XII. Communications

Note that the outline is a proposal, not a final standard. On the other hand, the BS 7750 is an existing and usable standard.
- April 6, 1992. Launching of BS 7750.

PROPOSED TRAINING

One characteristic of environmental management systems is that they are a continual process and therefore demand continuing appraisal, review, and training. The following is a recommended staff training syllabus for a BS 7750 with ISO 9000 as a base (1 day).

Introduction
    The global environment of business
    Regulations—environmental, product liability, health and safety
    Market demands
    Quality standards
    Our regulations
Concerns to our organization
    Our product and its use
    Our packaging
    ALL emissions
    Water discharges
    Material and energy use
    Solid and toxic waste
    Noise and nuisance
    Health and safety
    Landscape, wildlife, amenity
BS 7750
    The standard
    Relationship to ISO 9000
    The elements
    Costs and benefits
    What is involved
The environmental management system
    Our issues
    Defining, managing and controlling our issues
    The checklist
    Documentation
    Manual
    Procedures
    Instructions
    Forms

For additional training in the ISO series see Appendix C.

OVERVIEW OF THE STANDARD

The BS 7750 is under the direction of the Environment and Pollution Standards Policy Committee. It was developed in response to increasing concerns about environmental protection and performance. The standard in its entirety contains a management system for ensuring and demonstrating compliance with stated environmental policies and objectives. At this point the BS 7750 shares common management system principles with the ISO 9000. For a comparison of this sharing of principles see Table 9.1.

In addition to the stated specifications, the BS 7750 provides guidance for both the specifications and the implementation process of the environmental requirements.

The standard applies to all levels of environmental management performance as well as to particular sectors dealing with environmental issues. Particular sectors may be defined as industrial sectors having:

- Complex environmental effects on the society
- Large numbers of constituent companies
- Widely differing operations and disciplines
- Temporary and/or off-site activities
- Substantial use of subcontracting

It is important to recognize that even though the BS 7750 is a compliance standard, it does not in any shape or form grant immunity from any and all legal obligations.

Specifications for Environmental Management and Quality Assurance Standards

The aim of this section is to summarize each of the clauses in the standard and to provide the reader with an overview of what environmental quality is all about.

Specification. In this section the standard establishes the need for such a standard and provides an overview of it. In addition to this overview, the standard provides a minimal diagram for the implementation process recognizing that the system applies to all types and sizes of organization.

For an extensive model of the implementation process, the reader may want to review Chapter 5 on the implementation of ISO 9000. Since both standards are certifiable and have common elements, the process of implementation is the same. In a summary format, that implementation process follows the five-stage model shown in Figure 9.1.

Section 1: Scope. The standard specifies requirements for the development, implementation, and maintenance of an environmental management system. Its aim is to ensure compliance with a stated policy and objectives. In addition, the scope defines the applicability of the standard to any organization that:

- Assures itself of compliance with a stated policy
- Demonstrates such compliance to others

Section 2: Informative References. The guideline provides for the references used to define the environmental quality.

Section 3: Definitions. In this section the special and specific vocabulary for the standard is defined.

Section 4: Environmental Management System Requirements. In this section the standard provides for the following.

Section 4.1: Environmental Management System. This clause defines the environmental management system as a means of ensuring that the organization's activities conform to its environmental policy and objectives. It specifically identifies two minimum requirements:

- The preparation of a documented system with procedures and instructions
- The effective implementation of the system

Section 4.2: Environmental Policy. This clause mandates the need for an organizational policy that ensures:

- Relevancy of all activities to the environmental effects

- Communication and understanding of the policy in the entire organization
- Availability of the policy
- Incorporation of a statement that deals with continual improvement in environmental performance
- Setting and publication of environmental objectives

Section 4.3:   Organization and Personnel

Section 4.3.1:   Responsibility, Authority, and Resources.   This clause mandates the definition and documentation of the responsibility, authority, and interrelations of the appropriate personnel who manage, perform, verify, and need the organizational freedom to carry out all work affecting the environment.

Section 4.3.2:   Verification Resources and Personnel.   This clause requires the identification of all in-house verification requirements as well as adequate resources and the assignment of trained personnel for the verification activities.

Section 4.3.3:   Management Representative.   This clause mandates the appointment of a management representative who, regardless of other duties, has the defined responsibility and authority for ensuring the requirements of the standard.

Section 4.3.4:   Personnel, Communication, and Training.   This clause requires at least four items for compliance in the area of establishing and maintaining awareness for the employees of the organization. They are:

- Compliance with policy, objectives, procedures, and requirements of the standard
- Environmental effects of their work activities
- The individual role and responsibility in achieving compliance with the requirements
- The potential consequences of departure from the stated procedures

Section 4.4: Environmental Effects

Section 4.4.1: Register of Legislative, Regulatory, and Other Policy Requirements. This clause mandates that the organization must establish and maintain all appropriate legislative, regulatory, and policy requirements pertaining to the environmental aspects of all activities.

Section 4.4.2: Communications. This clause mandates the establishing and maintaining of procedures that affect all communications concerning environmental effects.

Section 4.4.3: Environmental Effects Evaluation and Register. This clause requires that the organization must establish and maintain procedures for examining and assessing the environmental effects. The procedures shall include, where appropriate, the following:

- Controlled and uncontrolled emissions
- Controlled and uncontrolled discharges to water
- Solid and other wastes
- Contamination of land
- Any use of natural resources
- Noise, odor, dust, vibration, and visual impact
- Any specific impact on the ecosystem

The emphasis of this clause is on the second part, which identifies the specificity of the procedures in reference to the consequences from the above. The standard identifies four concerns that must include effects arising, or likely to arise, as consequences. They are:

- Normal operations
- Abnormal conditions
- Incidents, accidents, and potential emergencies
- Past, current, and planned activities

Section 4.5: Environmental Objectives and Targets. This clause requires the organization to establish and maintain procedures that specify its environmental objectives. However, these procedures, objectives, and targets shall always be consistent with the environmental policy and shall quantify, whenever possible, the commitment to continual improvement.

Section 4.6: Environmental Management Program. This clause requires the organization to establish and maintain a program for achieving the objectives and targets. This program will include:

- Designation of responsibility for targets
- The means by which they are to be achieved

Furthermore, the clause of this standard requires the organization to establish and maintain separate programs relating to new developments and/or changed products. These separate programs are to define the:

- Environmental objectives
- Mechanisms for their achievement
- Procedures for dealing with changes
- Corrective mechanisms

Section 4.7: Environmental Management Manual and Documentation

Section 4.7.1: Manual. This clause mandates that the organization establish and maintain a manual to:

- Display the policy, objectives, targets, and programs
- Document the key roles and responsibilities
- Describe the interactions of system elements
- Provide direction to related documentation

Section 4.7.2: Documentation. This clause requires the organization to establish and maintain procedures for controlling all documents required by the standard and to ensure that:

- They can be identified by the appropriate organization, function, activity, and personnel
- They are periodically reviewed and or revised
- Current versions are available where applicable

- Obsolete documents are promptly removed from all points of issue and use

Section 4.8: Operational Control

Section 4.8.1: General. This clause requires that the management responsibilities must be defined to ensure that control, verification, measurement, and testing in the organization are adequate and effective.

Section 4.8.2: Control. This clause requires that the organization must identify all activities that affect or have the potential to affect the environment. These activities should be monitored to ensure that they are carried out under controlled conditions. To facilitate this insurance special attention should be given to:

- Documented work instructions
- Procedures dealing with purchasing or contracted activities
- Control of relevant process characteristics
- Approval for planned processes and equipment
- Criteria for performance

Section 4.8.3: Verification, Measurement and Testing. This clause requires that the organization must establish and maintain procedures for verification of compliance with specified requirements. In addition, the standard provides that each relevant activity should:

- Identify and document the verification
- Specify and document the verification
- Establish and document acceptance criteria
- Assess and document the validity of the verification

Section 4.8.4: Noncompliance and Corrective Action. This clause mandates the existence of investigative and corrective action plans in the event of noncompliance. Some guidelines are given for the direction of the investigative and corrective action plans. They are:

- Determine the cause
- Draw a plan of action

- Initiate preventive action
- Apply controls to ensure that the preventive action is effective
- Record any changes in procedures

Section 4.9: Environmental Management Records. This clause requires the organization to establish and maintain records to demonstrate compliance with the requirements of the environmental management system.

Section 4.10: Environmental Management Audits

Section 4.10.1: General. This clause requires the organization to establish and maintain procedures for audits. The focus of the audits according to the standards should be to identify whether:

- Conformity to the standard exists
- Implementation of the environmental system is effective
- The environmental system is fulfilling the environmental policy

Section 4.10.2: Audit Plan. This clause requires that the audit plan must deal with the following points:

- Specific activities, which will include:
    Organizational structures
    Administrative and operational procedures
    Work areas
    Documentation
    Environmental performance
- The frequency of the audit
- The responsibility for auditing each activity area
- The personnel requirements
    Independence
    Expertise
    Support
- The protocol for conducting the audits
- The procedures for reporting audit findings to those responsible.
    The reporting should cover the following:
        Conformity or nonconformity to the standards

The effectiveness of the implemented system
The effectiveness of the corrective actions
Conclusions and recommendations

Section 4.11: Environmental Management Reviews. This clause requires the organization to review at appropriate intervals the environmental system to ensure continuing suitability and effectiveness. The results of such review should be published if the organization has a commitment to do so.

Annexes (Appendixes). The standard provides some lengthy guidelines to fulfill the requirements. The four appendixes are not certifiable; however, they provide a good basis for understanding the standards.

Appendix I is the lengthiest and provides guidelines on how to develop the system requirements.
Appendix II provides a comparison between the BS 7750 and BS 5750.
Appendix III provides a comparison between the BS 7750 and the Eco-Audit Regulation (Version 3, December 1991).
Appendix IV provides the bibliography of the standard.

REFERENCES

BS 7750:1992. *Specification for Environmental Management Systems*, BSI, London, UK.
Rothery, B. (1993). *BS 7750*. Gower Press. Brookfield, VT.
Taguchi, G. (1987). *System of Experimental Design*, Vol. 1 and Vol. 2. UNIPUB/Kraus International Publications, White Plains, NY.

10
ISO and Other Quality Systems

The ISO system is a quality management system and is only one of many systems that different countries and/or industries utilize to assess their quality. This chapter compares the ISO with some of the most common systems.

As we have seen, the ISO standards were developed by a committee of quality professionals who are familiar and use quality system concepts in their jobs. These standards were in no way thought of or developed from scratch. In fact, they were developed by taking specific elements from other standards such as: MIL-I-45408, nuclear power plant regulations (appendix B to 10CFR50), MIL-Q-9858A, British standards (BS 5750), medical device manufacturer regulations (the Food and Drug Administration's Good Manufacturing Practices), Canadian standards (Z299), and other regulations. The ISO standards are indeed what Arter (1992) called "a gourmet soup."

Since the standard is a combination of other standards, it follows that there must be some common elements as well as differences. Let us look at some of these.

ISO AND TQM

At no other time have two quality systems (movements) existed simultaneously. Both ISO and total quality management (TQM) are very powerful systems with both similarities and differences. The similarities are self-explanatory but the differences are a little more convoluted. We will offer a brief description of the latter. For a detailed discussion, see Welch (1993), Harral and Berg (1993), and Finlay (1992).

Similarities.   The similarities between ISO and TQM pertain to quality philosophies, influence from other cultures, training and education, record keeping, customers and subcontractor relationships.

Differences.   The differences between ISO and TQM are more significant. They pertain to views of success, views of the process, views of responsibility, and views of flexibility.

Views of success. The ISO uses registration as the measure of success. TQM, on the other hand, uses customer satisfaction as the measure of success.

Views of the process. The ISO emphasizes controlling processes and replicability. TQM, on the other hand, emphasizes process improvements as well as controlling good processes.

Views of responsibility. In the ISO the responsibility is assigned to anyone who can have an effect on the quality of the product. Quite often, management assigns a representative. In general, the ISO guidelines indicate that responsibility should be assigned to individuals. TQM, on the other hand, emphasizes teamwork and involvement in implementing quality systems.

Views of flexibility. The ISO provides the structure, control, and adherence to procedures. As a consequence, it restricts the freedom an organization has to change and adapt. These attributes are also in opposition to the TQM philosophy of removing barriers that allow quality improvement changes to be made rapidly.

ISO AND DEMING-BASED TQM

ISO has clear requirements that may or may not be particularly significant in Deming-based TQM (DBTQM). ISO requirements dictate that contract review be addressed in very specific terms, while DBTQM leaves the details entirely up to the organization. Similar elements are the issues of design control, inspection, testing, and many more.

The basis of DBTQM is statistical understanding, yet ISO treats statistical process control as something of an afterthought.

ISO AND THE MBNQA

Finlay (1992, p. 1) suggests that a comparison between these two systems may be an "apples-versus-citrus-salad comparison." He continues the analogy by saying that the "ISO is like three starched white business shirts—small, medium, and large—form-fitting but not expected to cover the whole body." "MBNQA," on the other hand, "is like a giant, one-size-fits-all T-shirt with 33 pockets in which specific articles are to be placed."

It is indeed very difficult to compare the two systems. They are different. Their assumptions are different. The results are different. The evaluation is different. The MBNQA process is designed to recognize and award those firms with outstanding records of quality performance. The purpose of the program is very different from the purpose behind the development of the ISO criteria. While the use of the ISO may be a good starting point in establishing a quality system, the criteria used in evaluating candidates for the MBNQA are much more detailed and extend beyond those areas covered by the ISO.

The MBNQA criteria are results-oriented and cover all operations, processes, and work units of a company. The evaluation procedures emphasize the dynamics involved in the integration of all aspects of a firm's quality system and the firm's continuous improvements in quality (Breitenberg, 1993).

The ISO requirements are clearly defined, but the organization has the freedom to define "how" these requirements will be met. The ISO concentrates almost exclusively on results criteria, although process criteria may meet some ISO requirements—depending on the lead auditor.

The MBNQA guidelines are documentation-dependent. (The word "document" means *to state requirements on paper before an event occurs*. It is always *before* an action. A record, on the other hand, shows what happened. It is always *after* an action.) The guidelines are more results-oriented than process-oriented, but the organization is required to follow both results and process criteria. MBNQA requires specific organizational involvement and change.

ISO AND THE SHINGO PRIZE

The ISO has specific requirements and the evaluation is based on whether or not there is compliance to these requirements. The Shingo Prize, on the other hand, is a prize for excellence for U.S. companies that have demonstrated outstanding achievement in manufacturing processes, productivity improvement, quality enhancement and customer satisfaction.

One of the major differences between the two systems is that the application process for the Shingo Prize may be used as a vehicle for improvement.

ISO AND THE QS-9000 REQUIREMENT

In the automotive industry for the past 14 years everyone has been talking about quality. As a consequence, every major company focused on its own requirements, criteria, and evaluation methods to make sure that its product was better than the competition's. The result was to have product-oriented systems such as Ford's Q101, Chrysler's Pentastar, GM's Targets for Excellence, and many others.

The proliferation of all these standards created confusion between the supplier base. Everyone had an idea what the standards were or how could they be implemented, but in reality nobody really knew, and the quality system was really dependent on the knowledge and experience of the auditor.

That was true until the Automotive Task Force through the Automotive Industries Action Group (AIAG) got together to standardize some of the criteria, including evaluations as well as the content of the requirements. For the past several years they have been able to standardize the statistical process control (SPC), measurement systems (R&R studies), FMEA, control plans, and advanced quality planning.

So, what is the AIAG? and what are the expected benefits? The AIAG is a consortium of Ford, Chrysler, GM, Volvo—Heavy Truck, Navistar, PACCAR, Mack, and Freightliner. Their objective includes a continuing evolution to one set of quality system guidelines for both truck and automobile industries, minimizing exceptions (Fette, 1994). The envisioned benefits are standardization, consistency, and improved relations with suppliers.

As of September 1, 1994, the automotive industry through the AIAG's efforts published an official harmonized set of standards and called the system: *Quality System Requirements: QS-9000*. As already stated, the ISO is a quality management system. So the quality system developed by the AIAG begins with the ISO as the base. However, that base is expanded with a merged version of all the specific requirements of the Big Three's (Chrysler, Ford, GM) existing assurance standards. The result of this combination is a standard that exceeds the ISO. The areas that are beyond the traditional ISO are:

- Customer specific requirements
- Customer satisfaction
- Business planning
- Continuous improvement
- Cost of quality
- Health, safety, and environment
- Human resource development
- Failure mode and effect analysis
- Production part approval process
- Manufacturing capabilities
- Statistical process control
- Just-in-time delivery

In addition to these added common requirements, the AIAG provides for even more requirements for customer-dependent requirements, as they apply. For more detailed information on the QS-9000, see Stamatis, 1995, and AIAG, 1994.

ISO AND MIL-I-45208A

While the specific differences between ISO and MIL-I-45208A are numerous, the general difference between the two is that of scope. The military standard is concerned with conforming supplies through the use of inspection programs, and ISO is concerned with compliance and customer satisfaction (somewhat) through the use of quality management. An element-by-element comparison follows:

MIL-I-452208A paragraph		Q91
1. Scope 2. Applicable documents		Not applicable
3.1	Contractor responsibilities	4.3, 4.10.
3.2.1	Inspection and testing documentation and records	4.10, 4.12, 4.16
3.2.2		
3.2.3	Corrective action	4.14
3.2.4	Drawings and changes	4.5
3.3	Measuring and test equipment	4.11
3.4	Process controls	NSC
3.5	Indication of inspection status	4.12
3.6	Government-furnished material	4.7
3.7	Nonconforming material	4.13
3.8	Qualified products	NSC
3.9	Sampling inspection	4.10
3.10	Inspection provisions	NSC
3.11	Government inspection at subcontractor/vendor facilities	4.6.3, 4.6.4
3.12	Receiving inspection	4.10.2
3.13	Government evaluation	4.3

NSC = not specifically covered in Q91.
MIL-I-45208A contains fewer requirement than MIL-Q-9858A. Q91 paragraphs reference: 4.3, 4.5, 4.6, 4.7, 4.10, 4.11, 4.12, 4.13, 4.14.

ISO AND MIL-Q-9858A

For a comparison of the two standards we provide the following element-by-element comparison.

MIL-Q-9858A	Q91
1. Scope 2. Superseding, supplementation, and ordering	4.1
3. Quality program management	
3.1 Organization	4.1.2,
3.2 Initial quality planning	4.2.3[a]
3.3 Work instructions	4.2.2, 4.9, 4.16
3.4 Records	4.3.4, 4.5, 4.6.3, 4.9, 4.10.4, 4.16, 4.14
3.5 Corrective action	4.14
3.16 Costs related to quality	NSC (see Q94, 6)
4. Facilities and standards	
4.1 Drawing, documentation, and changes	4.4, 4.4.1–4.4.9
4.2 Measuring and testing equipment	4.11
4.3 Production tooling used as media inspection	4.11
4.4 Use of contractor's inspection equipment	Not specifically covered
4.5 Advanced metrology requirements	Not specifically covered
5. Control of purchases	
5.1 Responsibility	4.6.1–4.6.4, 4.10.2
5.2 Purchasing data	4.6.3[a]
6. Manufacturing control	
6.1 Materials and materials control	4.10.1–4.10.5
6.2 Production processing and fabrication	4.9, 4.10.2, 4.10.3, 4.14
6.3 Completed item inspection and testing	4.10.4
6.4 Handling, storage, and delivery	4.15.2, 4.15.3, 4.15.6
6.5 Nonconforming material	4.13, 4.13.2
6.6 Statistical quality control and analysis	4.20
6.7 Indication of inspection status	4.10.3–4.10.4
7. Coordinated government/contractor action	Not applicable

[a]9858A has more requirements than Q91.

There is little or no provision in MIL-Q-9858A for (a) document control, (b) product traceability, (c) calibration status of inspection and test equipment, or (d) internal quality audits. Training and servicing are also not explicitly cited.

ISO AND FDA (GMP)

The Good Manufacturing Practices (GMP) provide some basic quality assurance methods and practices common to ISO. However, in some cases they go further in their quest to define specific areas that the ISO does not address at all. An element-by-element comparison follows (21 CFR Part 8.20, *Federal Register*, Vol. 43, Vol, 41, p. 31527):

GMP		Q91
Subpart A	General provisions	
820.1	Scope	1
.3	Definitions	3
.5	Quality assurance program	4
Subpart B	Organization and personnel	
820.20	Organization	4.1.2
	General	4.1, 4.2
	Quality assurance program requirements	4.2
	Audit requirements	4.1.3, 4.17
.25	Personnel	
	Training	4.18
	Health and cleanliness	Not covered
Subpart C	Building	
820.40	Building	Implied 4.9
.46	Environmental control	Implied 4.9
.56	Cleaning and sanitation	Not covered
Subpart D	Equipment	
820.60	Equipment	
	General	4.9
	A–D	Not covered
.61	Measurement equipment	4.11
Subpart E	Control of components	
820.80	Control of components	
	General	4.10.1, 4.13
	Acceptance	4.10.1, 4.10.2–4.10.4
	Storage and handling	4.15.3, 4.15.2
.81	Critical devices	
	Acceptance	4.10, 4.20
	Supplier agreement	Implied 4.3

GMP		Q91
Subpart F	Production and process controls	
820.100	General	4.9
	Specification control	4.9
	Changes	4.9
	Processing controls	4.9
.101	Critical devices (re-100)	
.115	Reprocessing of devices and components	4.13
.116	Critical devices (re-115)	
Subpart G	Packaging and labeling control	
820.120	Device labeling	4.15.4
.121	Critical devices (re-120)	
.130	Devices packaging	4.15.4
Subpart H	Holding, distribution, and installation	
820.150	Distribution	4.15.2
.151	Critical devices (re-150)	
.152	Installation	Implied 4.19
Subpart I	Device Evaluation	
820.160	Finished device inspection	4.10.4
.161	Critical devices (re-160)	
.162	Failure investigating	4.14.2
Subpart J	Records	
820.180	General	4.16
	Confidentiality	Not covered
.181	Device master record	4.5.2
.182	Critical devices (re-181)	
.183	Device history record	4.8, 4.16
.185	Critical device (re-184)	
.195	Critical devices	
	Automated data processing	Not covered
.198	Complaint files	4.14.2

In general, the FDA GMP regulations are much more detailed than Q91; however, the latter are much broader in defining the quality system. The requirements for critical devices are very detailed. The GMP's do not speak to the following Q91 paragraphs: 4.3, Contract review; 4.4, Design; 4.7, Purchase supplied product; 4.12, Inspection and test status; 4.19, Servicing.

Table 10.1 ISO 9000, SAE, and CASE Quality System Requirements: Cross-Reference Matrix

ISO 9001 ANSI/ASQC Q91-1994		ISO 9002 Q92-1994	ISO 9003 Q93 1994	AS 7201 10/1/91	AS 7202 10/1/91	CASE-5A 11/15/91
1	Scope	1	1	1.1, 5.4.1	1.1, 1.4	1.A–G, 2.F
2	Normative reference	2	2	2.1	2.1	
3	Definitions	3	3	1.3	1.3	
3.1	Product	3.1	3.1			
3.2	Tender	3.2	3.2			
3.3	Contract; accepted order	3.3	3.3			
4	Quality system responsibility	4	4	N/A	N/A	N/A
4.1	Management responsibility	4.1	4.1	N/A	N/A	N/A
4.1.1	Quality policy	4.1.1	4.1.1			2.B
4.1.2	Organization	4.1.2	4.1.2	N/A	N/A	N/A
4.1.2.1	Responsibility and authority	4.1.2.1	4.1.2.1	3.2.1	3.2.1	2.D, 8.A, 8.B
4.1.2.2	Resources	4.1.2.2	4.1.2.2			2.C
4.1.2.3	Management representative	4.1.2.3	4.1.2.3			
4.1.3	Management review	4.1.3	4.1.3			
4.2	Quality system	4.2	4.2			2.A
4.2.1	General	4.2.1	4.2.1			
4.2.2	Quality-system procedures	4.2.2	4.2.2			
4.2.3	Quality planning	4.2.3	4.2.3			
4.3	Contract review	4.3	4.3			
4.3.1	General	4.3.1	4.3.1			
4.3.2	Review	4.3.2	4.3.2			

4.3.3	Amendment to contract	4.3.3				
4.3.4	Records	4.3.4	4.3.4			
4.4	Design control			N/A	N/A	N/A
4.4.1	General					
4.4.2	Design and development planning					
4.4.3	Organizational and technical interfaces					
4.4.4	Design input					
4.4.5	Design output					
4.4.6	Design review					
4.4.7	Design verification					
4.4.8	Design validation					
4.4.9	Design changes					
4.5	Document and data control	4.5		N/A	N/A	N/A
4.5.1	General	4.5.1				
4.5.2	Document and data approval and issue	4.5.2				3.B, 5.0
4.5.3	Document and data changes	4.5.3				
4.6	Purchasing	4.6		N/A	N/A	N/A
4.6.1	General	4.6.1				9.A
4.6.2	Evaluation of subcontractors	4.6.2		3.1.2	3.1.2	9.C
4.6.3	Purchasing data	4.6.3		3.1.1, 3.2.1	3.1.1, 3.2.1	6.D, 9.B
4.6.4	Verification of purchased product	4.6.4				

Table 10.1 (Continued)

ISO 9001 ANSI/ASQC Q91-1994		ISO 9002 Q92-1994	ISO 9003 Q93 1994	AS 7201 10/1/91	AS 7202 10/1/91	CASE-5A 11/15/91
4.6.4.1	Supplier verification at subcontractor's premises	4.6.4.1				
4.6.4.2	Customer verification of subcontracted product	4.6.4.2				
4.7	Control of customer-supplied product	4.7	4.7			
4.8	Product identification and traceability	4.8	4.8	3.2.2, 5.3	3.2.2, 5.3	3.C, 6.D, 10.1
4.9	Process control	4.9	4.9	4.0	4.0	
4.10	Inspection and testing	4.10	4.10	N/A	N/A	N/A
4.10.1	General	4.10.1	4.10.1			
4.10.2	Receiving inspection and testing	4.10.2	4.10.2	3.7	3.7	3.A–D
4.10.2.1		4.10.2.1				
4.10.2.2		4.10.2.2				
4.10.2.3		4.10.2.3				
4.10.3	In-process inspection and testing	4.10.3				
4.10.4	Final inspection and testing	4.10.4		3.2.2, 3.3, 3.4.2, 5.2	3.2.2, 3.3, 4.4.2, 5.2	
4.10.5	Inspection and test records	4.10.5		3.3	3.3	6.A
4.11	Control of inspection, measuring, and test equipment	4.11	4.11	3.9	3.9	4.A–D
4.11.1	General	4.11.1	4.11.1			

4.11.2	Control procedures	4.11.2			
4.12	Inspection and test status	4.12	3.2.2	3.2.2	
4.13	Control of nonconforming product	4.13	3.5	3.5	3.B, 3.E
4.13.1	General				
4.13.2	Review and desposition of nonconforming product				
4.14	Corrective and preventive action	4.14			
4.14.1	General				
4.14.2	Corrective action		3.6	3.6	2.E, 10.B
4.14.3	Preventive action				
4.15	Handling, storage, packaging, preservation, and delivery	4.15	N/A	N/A	N/A
4.15.1	General	4.15.1	3.3	3.3	
4.15.2	Handling	4.15.2	3.4.1	3.4.1	10.A, 10.I
4.15.3	Storage	4.15.3	3.4.1	3.4.1	7.10.A, 10.H, 11
4.15.4	Packaging	4.14.4	3.2.2, 3.3, 3.4.1, 3.4.2, 5.3	3.2.2, 3.3, 3.4.1, 3.4.2, 5.3	3.A, 10.A, 10.C–J
4.15.5	Preservation	4.15.5			
4.15.6	Delivery	4.15.6	3.4.1	3.41	10.G
4.16	Control of quality records	4.16	3.4.2, 5.1	3.4.2, 5.1	3.B, 3.D, 6.A–C, 8.C, 10.D
4.17	Internal quality audits	4.17			
4.18	Training	4.18	3.8	3.8	8.A–C
4.19	Servicing	4.19			
4.20	Statistical techniques	4.20			
4.20.1	Identification of need				
4.20.2	Procedures				

ISO AND SAE 7201-2 AND CASE-5A

The standards AS 7201 and AS 7202 are recommended by the Society of Automotive Engineers (SAE). The standards for the airline and aerospace industry organization are recommended by the Coordinating Agency for Supplier Evaluation (CASE-5A). CASE-5A is an airline and aerospace industry organization that publishes a register of evaluated suppliers approved for or qualified to government, regulatory, and contracting organizations requirements and specifications.

Table 10.1 shows the relationship of ISO, AS 7201-2 and CASE-5A. The table is intended as a quick reference guide to the system requirements. N/A signifies that there is no required cross-reference to the specific section of ISO 9001. A blank space indicates that there is no direct requirement in that specification that corresponds to the ISO 9001.

A warning to the reader. Some references to the ISO 9000 series and SAE requirements are open to some individual discretion.

REFERENCES

Arter, D. R. (November 1992). "Demystifying the ISO 9000.Q90 series standards." *Quality Progress*.

Automotive Industries Action Group (AIAG) (1994). *Quality System Requirements: QS-9000*. AIAG, Southfield, MI.

Breitenberg, M. (April 1993). *ISO 9000: Questions and Answers on Quality, the ISO 9000 Standard Series, Quality System Registration, and Related Issues*. U.S. Department of Commerce. Publication number NISTIR 4721.

Fette, D. V. (January 1994). "Industry standards reports—Automotive." *The EC Report on Industry*.

Finlay, J. S. (August 1992). "ISO 9000, Malcolm Baldrige award guidelines and Deming/SPC-based TQM—A comparison." *Quality Systems Update*.

Harral, W. M., and Berg, D. L. (Fall 1993). "Implementing TQM in an ISO framework." *Automotive Division Newsletter*.

Stamatis, D. H. (1995). *Integrating QS-9000 with Your Automotive Quality System*. Quality Press, Milwaukee, WI.

Welch, C. (November 1993). "ISO 9000 and TQM: Finding balance." *Proceedings: 1st Annual International Conference on ISO 9000*. Lake Buena Vista, FL.

11

The Future of the ISO

Predicting the future of anything is risky at best. Predicting what the world holds in store for quality is even more difficult. Our attempt, however, is going to be more generic than specific, based on the trends of the last couple years.

According to a survey by Mobil Corp., reported in the April 1994 issue of *ONQ*, the total number of registrations to the ISO 9000 standards grew by 70% during the first 9 months of 1993. Registrations grew to 45,000 in October 1993 from a year-end 1992 total of 26,400.

The same survey reported that the United Kingdom has been issued the greatest number of registrations, with 62.5% share; the rest of Europe follows with a 21.5% share, Australia and New Zealand 7.1%, North America 4.7%, and the far East 3.5%.

The survey further reported that the number of countries in which registrations were issued rose from 48 to 60—a 25% increase. The most dramatic increase in the number of registrations—over 150%—occurred in Japan and the United States. Singapore and Malaysia followed with about a 100% increase, with Germany close behind.

The results of the survey indicate that the ISO is a world system and no longer the quality system only for the European Union. In fact, steps are being taken to create an international system for the recognition of ISO certificates. Towards that goal, the International Organization for Standardization and the International Electrotechnical Commission are being asked to set an ad hoc committee to develop the rules and guidelines for the system, define the responsibilities for its governing body, outline its governing body, outline its operating method, and set a timetable for its establishment (*ISO 9000 News*, July/August 1993).

VISION 2000

The International Organization for Standardization continues to develop industry-specific standards through TC 176 to avoid proliferation of individual standards throughout the world. The vision is to have a truly global marketplace by having an internationally accepted mechanism that everyone endorses.

The goals of the vision 2000 may be categorized in the following areas:

Universal acceptance

- The standards to be widely adopted and used worldwide
- Few complaints from users in proportion to their use
- Few sector-specific supplementary standards are being used or developed

Current compatibility

- Supplements to existing standards do not change or conflict with requirements in the parent standard
- The numbering and clause structure of a supplement remains the same as that of the parent document
- Supplements are not "stand alone" documents, rather they are to be used with the parent document

Forward compatibility

- Revisions affecting requirements in existing standards are few in number and minor or narrow in scope

- Revisions are accepted for new and existing contracts

Forward flexibility

- Supplements are few in number, but can be combined as needed to meet the needs of virtually any industry or economic sector or generic category of products
- Supplements or addendum architecture allows new features or requirements to be consolidated into the parent document at a subsequent revision if the supplement's provisions are found to be used (practically) universally

The essence then, of the vision 2000 is to have standards that meet market needs. There is no value in a standard that is not wanted or used in the marketplace—for whatever reason.

The plans for the vision 2000 documents as of this writing are:

Planned for 1996
Non-industry/economic-sector specific
Product-specific documents containing technical requirements for product test (to be developed within industry/economic sectors)

ISO 9000 SPREADS

The ISO 9000 answers many of the quality deficiencies and irregularities that many organizations are forced to suffer. It establishes the foundations of a quality management system and provides a consistent quality management system for any organization committed to quality. The ISO allows for the expansion of that system through continuous improvement to reach a level of world class.

The ISO recognizes that quality is a journey, not the destination. As a consequence, a continual improvement defines a world-class, mature, quality organization as one in which an integrated philosophy of improving productivity, quality of product, and quality of work life for all stateholders is present. The intent for such organizations is to demonstrate consistent, measurable, and positive results (Scheffler, 1993).

The ISO standards are so fundamental to the overall quality that the ISO Committee is expanding its scope to specific industries through specific directives and guidelines. The directives have a focus to standardize the regulations across different countries where the guidelines provide hints and recommendations as to *how* to go about implementing the standard in a given organization.

Many directives are under consideration. Some of the most important are (*ISO 9000 News*, July/August 1993; *ONQ*, 1994; Kolka, 1994):

Machinery safety integration: This is the basic approach of the directive. Safety is to be integrated at the design stage and maintained throughout the foreseeable lifetime of the machine.

Telecommunications conformity assessment: The basic tenet of this directive is on the conformity requirements and local representation.

Software: The objectives of this directive are to ensure an adequate level of protection for those who create computer software and to promote the free circulation of computer software within the community. While the objectives seem to be very straightforward, there are some concerns. They are:

- Interpretation of the ISO 9000-3 as a guide. In this book we have attempted to give an interpretation and focus for proper implementation of the ISO 9001 in the software industry.
- Scope of the software sector scheme, i.e., the TickIT and the SQSR program.
- Auditor qualification and certification process.
- National accreditation process and recognition of accreditations.
- Obtaining input from the customer.
- Information technology (IT) registration logo.

Medical devices: The European community has been developing three medical devices directives. The first, the active implantable medical device, serves as the parent directive for the structure of the general medical devices directive and the in vitro diagnostics directive, which is about to move from a working draft to a proposal.

Product safety directive: This directive, which went into effect on June 29, 1994, is aimed at the rapid exchange of information in the event of serious risk to the health and safety of consumers.

Packaging: The packaging and packaging waste directive covers

ALL industrial, commercial, and household, primary (sales packaging), secondary (group packaging), and tertiary (transport packaging). Specifically, it defines packaging in all forms, prevention recovery, recycling, and disposal. Furthermore, it sets targets regarding the recovery of packaging waste at no later than 5 years—except for Greece, Ireland, and Portugal, which are allowed until December 31, 2005, with at least 25% recovery in the first 5 years.

EN 46000: This standard, which is to supplement, maybe even replace, ISO 9001 for the medical device industry, will embrace all the requirements of good manufacturing practice (GMP).

Health care TC forming: The International Organization for Standardization is creating a technical committee (TC) to prepare horizontal standards for the application of the ISO 9000 series of standards to health care products and for fundamental requirements for health care products. A private hospital in the United Kingdom has achieved registration to ISO 9002, illustrating the relevance of the series to an increasingly wide range of organizations (*ISO 9000 News*, September/October 1993).

In addition to the preceding areas the ISO TC 176, as of this writing, is working on development of the following:

* Draft international standard (DIS) 9004-3: Quality management and quality system elements—Part 3: Guidelines for processed materials (expected to be out at any time).
* Draft international standard (DIS) 9002: Quality management and quality system standards—Part 2: Generic guidelines for the application of the certifiable standards—ISO 9001, 9002, and 9003 (expected to be out at any time).
* Draft international standard (DIS) 9004: Quality management and quality system elements—Part 4: Guidelines for quality improvement (expected to be out at any time).
* Committee draft (CD) 9004-6: Quality management and quality system elements—Part 6: Guidelines quality plans.
* Working draft (WD) 9004-5: Quality management and quality system elements—Part 5: Guide to quality assurance for project management.

- Working draft (WD) 10012-2: Quality assurance requirements for measuring equipment—Part 2: Measuring assurance.
- Committee draft (CD) 10013: Guidelines for developing quality manuals (expected to be out at any time).
- Working draft (WD) 10014: Guide to the economic effects of quality.
- Working draft (WD) 10015: Continuing education and training guidelines.

Perhaps one of the important areas where the ISO may have the greatest impact is the formation of ISO/TC 207. The mission of the Technical Committee (TC) 207 is to develop an international environmental management standard. Block (1994)suggests that this initiative reflects the recommendation of the Strategic Advisory Group on Environment (SAGE), an ad hoc committee created by ISO in 1991 to explore the feasibility of creating an environmental management standard similar to the ISO 9000 series of standards for quality management systems. (For a more detailed discussion of the environmental issues see Chapter 9.)

WORLDWIDE ACCEPTANCE

The Quality Expo International Conference in Chicago, Illinois, during April 1993 heard from international speakers how the ISO 9000 standard is gaining acceptance around the world (Kochan, 1993). That in itself is an excellent development. However, it is certainly not the end of the story.

Unless international harmonization and mutual recognition also are achieved, ISO 9000 will fail in its purpose of removing market barriers. Registrations, conformity assessments, and registration procedures must be recognized by all. This is far from the current reality, but a number of national and international bodies are beginning to tackle the issue.

While in the short run there may be some difficulties of recognition, as of this writing all key indicators show that the ISO standards are not a fad. They are the world standards of the future. Together with the directives and guidelines, they offer any organization the standardization and minimum consistency of a quality management system, second to none.

To be sure, individual industries may need to modify and/or add to the standards. However, the basic structure of the ISO is indeed universal and applicable to all industries. The additions and/or modifications may deal with the customer-specific requirements.

It is with regard to this consistency and standardization throughout the world of business that the ISO has the most profound effect. By having a common denominator of a quality system, organizations can make sound decisions on productivity, costs, and profitability based on sound comparisons.

The future is indeed very bright for the ISO. As time goes on, the difficulties, the uncertainties, the points of conflict within the standards will be addressed and in some cases removed or at least modified. To expedite this resolution the Technical Committees will strive to issue guidelines and directives for more uniformity.

REFERENCES

Anon. (July/August 1993). "US proposes committee on ISO 9000 and health care products." *ISO 9000 News*.

Anon. (September/October 1993). "Hospital gains ISO 9000 registration." *ISO 9000 News*.

Anon. (April 1994). "Registration to ISO 9000 grows worldwide." *ONQ*.

Block, M. R. (March 1994). "ISO/TC 207." *The European Marketing Guide*.

Kochan, A. (October 1993). "ISO 9000: Creating a global standardization process." *Quality*.

Kolka, J. W. (March 1994). "EN 46000 and the EU medical devices." *The European Report on Industry*.

Scheffler, S. (August/September 1993). "What is world class, mature quality?" *Continuous Journey*.

Appendix A
Unofficial Glossary of Acronyms

In this appendix we provide the reader with some of the most common acronyms in the world of the ISO. By no means is this an exhaustive list. Rather, it is a list that will facilitate the understanding of the acronyms most often found in the literature. We give the official name with some explanation of the term.

ANSI   **American National Standards Institute**
             Not a standards-writing body. Assures that member organizations that *do* write standards follow rules of consensus and broad participation by interested parties. ANSI is the U.S. member of ISO.

ASQC   **American Society for Quality Control**
             A technical society of over 100,000 quality professionals. Individual members throughout the world, but primarily for the United States. Publishes quality related

literature and the American National Standards. ASQC is the sole owner of RAB.

BSI   **British Standards Institute**
United Kingdom's standards writing body.

BSIQA   **BSI Quality Assurance**
One of (at present) 15 accredited certification bodies (registrars) in the United Kingdom. Assesses suppliers for conformance to the appropriate ISO 9000 series standards; registers those that conform. Organizationally separate from BSI.

CEN   **European Committee for Standardization**
Publishes regional standards (for EC and EFTA) covering nonelectrical, nonelectronic subject fields. (*See also* CENELEC)

CENELEC   **European Committee for Electrotechnical Standardization**
Publishes regional standards (for EC and EFTA) covering electrical/electronic subject fields. (*See also* CEN)

DIN   **Deutsche Institut für Normung**
Germany's standards-writing body.

DQS   **Deutsche Gesellschaft zur Zertifizierung von Qualitatssicherungs—systemen mbH (German Association for Certification of Quality Systems)**
A German registrar. Assesses suppliers for conformance to the appropriate ISO 9000 series standards; registers those that conform.

EAC   **European Committee for Accreditation of Certification Bodies**
A newly formed organization whose members represent the accreditation organizations of all EC and EFTA nations. It appears that it aspires to designation as a Specialized Committee for accreditation in the EOTC. Its mission will likely be to harmonize rules for accredita-

tion; facilitate mutual recognition of accreditation; provide advice and counsel to other committees in the EOTC framework on matters related to accreditation.

EC   **European Community**
See EU.

EEC   **European Economic Community**
Along with the European Atomic Energy Community (EURATOM) and the European Coal and Steel Community (ECSC), EEC was merged into the EC in 1967.

EFQM   **European Federation for Quality Management**
An organization of upper-level managers concerned with quality.

EFTA   **European Free Trade Association**
A framework within which its members strive to remove import duties, quotas, and other obstacles to trade and to uphold liberal, nondiscriminatory practices in world trade. (Similar to the goals of the old EEC.) Current members are Austria, Finland, Iceland, Norway, Sweden, Switzerland.

EOQ   **European Organization for Quality**
Formerly EOQC—European Organization for Quality Control. An independent organization whose mission is to improve quality and reliability of goods and services primarily through publications and conferences/seminars. Members are quality-related organizations from countries throughout Europe including Eastern-bloc countries. ASQC is an affiliate society member.

EOTC   **European Organization for Testing and Certification**
Set up by the EC and EFTA to focus on conformity assessment issues in the nonregulated spheres. Its purpose is to facilitate the development of certification and registration systems and the development of mutual recognition agreements.

EQNET **European Network for Quality System Assessment and Certification**
A business arrangement among quality system certification bodies (registrars). A single, major registrar per country is allowed. Its mission is to simplify the obtaining of quality system certificates in the several countries in which a multi-national supplier operates. The member registrars will issue several certificates simultaneously after performing a joint audit.

EQS **European Committee for Quality System Assessment and Certification**
EQS has been formed with the expectation that it will be designated as the Specialized Committee for quality systems in the EOTC. Its members are individual experts or delegations, not organizations. Members are appointed by each EC and EFTA country. The function of EQS is to: harmonize rules for quality system assessment and certification (registration); facilitate mutual recognitions of registrations; provide advice and counsel to other committees in the EOTC framework on matters related to quality system assessment and certification.

ETL **ETL Testing Laboratories**
A U.S. product testing and certification organization that has recently entered the quality system registration field.

EU **European Union**
A framework within which its members have agreed to integrate their economies and eventually to form a political union. Current member are Belgium, Denmark, France, Germany, Greece, Ireland, Italy, Luxembourg, The Netherlands, Portugal, Spain, United Kingdom, and Austria. Referred to as the European Community until 1994.

EUROLAB Has been formed with the expectation that it will be designated as the Specialized Committee for testing in the EOTC. The function of EUROLAB is to: provide an interface between the testing community and other concerned parties; accelerate development and harmonization of test methods; promote mutual acceptance of test results; provide to the EOTC expertise in the field of testing.

IEC **International Electrotechnical Commission**
A worldwide organization that produces standards in the electrical and electronic fields. Members are the national committees, composed of representatives of the various organizations that deal with electrical and/or electronic standardization in each country. Formed in 1906.

IQA **Institute for Quality Assurance**
British organization of quality professionals. Operates a widely recognized system of certification of auditors for quality systems.

ISO **International Organization for Standardization**
A worldwide federation of national standards bodies (87 at present). Produces standards in all fields except electrical and electronic, which are covered by IEC. Formed in 1947.

LRQA **Lloyd's Register Quality Assurance**
One of (at present) 15 accredited certification bodies (registrars) in the United Kingdom. Assesses suppliers for conformance to the appropriate ISO 9000 series standards; registers those that conform. Organizationally separate from Lloyd's Register Shipping.

MOU **Memorandum of Understanding**
A written agreement among a number or organizations covering specific activities of common interest. EQS was established with an MOU, as was EQNET. There are a number of MOUs covering mutual recognition of quality system registrations in which one of the signatories is a non-European registrar.

NACCB   **National Accreditation Council for Certification Bodies**
The British authority for recognizing the competence and reliability of organizations that perform third-party certification of products and/or registration of quality systems. Formed in 1984, it is the world's second such organization.

QMI   **Quality Management Institute**
The Canadian registrar of quality systems. Part of the Canadian Standards Association (CSA).

QMI   **Quality Management International**
A British consultancy in the field of quality.

RAB   **Registrar Accreditation Board**
A U.S. organization whose mission is to recognize the competence and reliability of registrars of quality systems and to achieve international recognition of registrations issued by accredited registrars. A subsidiary of ASQC.

RvC   **Raad voor de Certificatie (Dutch Council for Certification)**
The Dutch authority for recognizing the competence and reliability of organizations that form third-party certification of products, accreditation of laboratories, and/or registration of quality systems. Form in 1980, this was the first such organization.

UL   **Underwriters Laboratories**
A U.S. product testing and certification organization that has recently entered the quality system registration field.

UNI-CEI   **Ente Nazionale Italiano di Unificazione**
The Italian authority for recognizing the competence and reliability of organizations that perform third-party certification of products and/or registration of quality systems. A recent addition to the short list of accreditation organizations.

Appendix B

Critical Characteristics
for Procedures

This appendix provides a summary of the critical characteristics of all procedures and a generic example of procedure construction that meets the requirements for most quality systems, including the ISO 9000 standards.

Purpose: Why?   This section usually describes the reasons for writing the procedure. Most often the reasons are to establish the requirements, responsibilities, and methods for conducting some activity.

Scope: Where?   What departments, divisions, groups, individuals, etc., are expected to abide by the guidelines set out within the procedure, or if only part of a subject is being covered.

Responsibility: Who?   Specifically, who or what positions or departments will be expected to: maintain related documentation, conduct the various activities, follow up on findings and make reports.

Definitions: Provide the meaning of acronyms, not what the initials stand for but the actual definitions of the acronyms. Define every term that appears to be peculiar to the procedure that you are writing. Don't assume the reader knows anything.

Associated documents: Include descriptions of any forms and or documents used by the procedure itself. Instructions for forms: When a form exists, there must be a procedure for its use and instructions for completing the form. Identify every user block on the form with a letter or a number.

Procedures: This item provides the actual instructions. One cannot write a procedure unless s/he knows it will work. One will not know if it works unless one can do it himself or herself.

Audit statement: Briefly defines who, what, where, why, and when or how often the procedure will be audited.

Audit checklist: An itemized checklist that reflects the actual requirements and expectations of the procedure itself. Write the checklist plain enough for the reader to understand what is expected.

Note that not all these characteristics may be appropriate for all procedures. In fact, in some cases more may be appropriate. Use them when appropriate and applicable.

EXAMPLE 1: A GENERIC INSTRUCTION FORMAT

Reference:
Page: _____ of _____
Revision:
Revised Date:
Original Date:
Approval:
Title:

Work Instruction

1.0   Objective: XXXXXXXXXXXX
                      XXXXXXXXXXXX

2.0   Materials/   XXXXXXXXXXXX
       Tools   XXXXXXXXXXXX
                     XXXXXXXXXXXX

3.0
Procedure
In this section the description and/or the outline of each step of the instruction is written.

Step-by-Step Action

3.1   Step 1

3.2   Step 2

3.3   Step 3

3.4   Step 4

EXAMPLE 2:   ON-HIRE PROCESS

Title: On-Hire process	Authorised by: DHS	Release date: 5/6
Company: ABC, Inc.	Revision No. 0	Page 1 of 1
Responsibility: Operations	Replaces Procedure No ____ page ____ of ___	

Purpose:

To describe the process of providing a customer with the most suitable tank to meet their needs while satisfying internal and external requirements.

Procedure:

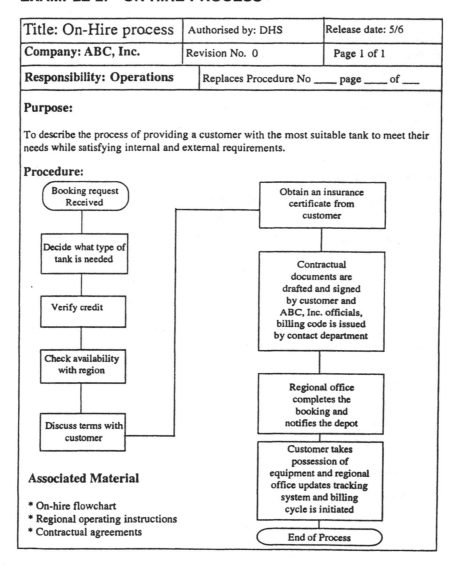

Associated Material

* On-hire flowchart
* Regional operating instructions
* Contractual agreements

Appendix C
Training Curriculum

In Chapter 5 we identified training as an indispensable characteristic of the implementation process. Because of the importance of training, we are identifying the curriculum for an appropriate ISO implementation. The curriculum is not a prescription for every organization; rather it provides the sequence for appropriate training. Each organization has to assess its own environment and then plan accordingly.

We have tried to give generic course outlines for the essential characteristics of any ISO implementation endeavor, but we recognize that some organizations may need more training in other areas as well. For example: project management, teams, employee empowerment, general quality training, specific quality tools (brainstorming, cause-and-effect diagram, process flow chart, FMEA, DOE, QFD, etc.).

A GUIDE FOR AN EXECUTIVE GENERIC OVERVIEW OF ISO 9000

This workshop is intended as an executive generic overview and a general review of the ISO. Specifically, it may serve a variety of industries, including: automotive, electronics, plastics, steel, printing, paper, and many more. It may be followed by specific training in the areas of implementation strategies, writing the documentation, and auditing training.

    I.   Overview
        A.   What are the standards?
        B.   Product life cycle wheel
   II.   Why standards
        A.   European Community
        B.   Liability directives
        C.   Safety directives
        D.   Market demands
  III.   Historical perspective
  IV.   Future directions
        A.   Vision 2000
        B.   A single quality management system
    V.   ISO 9000 acceptance
        A.   European Community
        B.   The United States
        C.   Worldwide
  VI.   ISO 9000 as a platform
        A.   Overview of the clauses in ISO 9001
        B.   Your suppliers and their role
 VII.   ISO and other systems
        A.   Quality management versus TQM
        B.   Malcolm Baldrige versus ISO
        C.   Mil-Q-9858A versus ISO
        D.   General conclusions
VIII.   Benefits
        A.   Question of value
        B.   Market and customer requirements

    IX. ISO structure
- A. Overview of the ISO structure
- B. Definitions
- C. Overview of the quality management system

    X. Synopsis of the ISO concept
- A. Implementation strategies
- B. Overview of documentation need and structure

A GUIDE FOR THE SERVICE INDUSTRY

This workshop is intended as an executive overview and a general review of the ISO in the service industry. It may be followed by specific training in the areas of implementation strategies, writing the documentation, and auditing training.

    I. Service quality
- A. Quality defined
- B. Product versus service
- C. What is service?
- D. Quality in service
- E. The ISO 9000 standard and services

    II. ISO 9000 overview
- A. History of ISO
- B. Structure of ISO
- C. Theory of ISO
- D. Overview of the individual clauses

    III. The contents of ISO 9001 and ISO 9004-2 explained

    IV. Quality system implementation
- A. Identify your business needs
- B. Obtain management commitment
- C. Organize training awareness
- D. Appoint a management representative
- E. Define project structure
- F. Prepare implementation functional areas

    V. Establish a management steering committee

    VI. Document the system
- A. Define the documentation needed

B. Identify the driving forces for completing the documentation
C. Identify the required documentation: formal and informal
D. Define the most appropriate hierarchy for your documentation
    1. Quality manual
    2. Quality procedures
    3. Work instructions
    4. Forms and tags
VII. Consolidation of people, processes, and documentation
    A. Understand the ISO standard in relation to the philosophy of "continual improvement"
    B. Train in audit system effectiveness
    C. System to report to steering committee
    D. Appropriate ALL necessary resources for training
    E. Conduct a preassessment
VIII. Registration strategy
    A. Initial contacts
    B. Initial response
    C. Lead time
    D. Preassessment services
    E. Costs
    F. Develop a relationship (your organization with the registrar)
    G. Be prepared for questions and more questions
IX. Register your company
    A. Define the scope of registration
    B. Prepare for the site audit
    C. The registration audit
    D. Audit management

ISO 9000-3 AND APPLYING ISO 9001 TO SOFTWARE DEVELOPMENT

Many U.S. companies are now manufacturing products that also include software as a portion of the deliverable item. As the increasing number of manufacturers consider ISO 9000 as an important step in their quality progress, the aspect of software as a product becomes more of an issue. All of those manufacturers, as well as those companies whose only product is software, will be faced with the realization that their software development process needs to be controlled if they expect to have their facility registered to the ISO 9001 standard.

With the ISO 9000 standards being written in the terminology of a manufacturing environment, ISO 9001 can become somewhat of a mystery to those people who are involved in software development. This course is designed to bridge the vocabulary gap between ISO 9000-3 and ISO 9001. It provides a comprehensive, yet simple, understanding of ISO 9000-3 and ISO 9001 for those who develop software. It applies to both the companies that only develop software as well as companies that include software as a portion of a delivered product or service.

This workshop is intended as an executive overview and a general review of the ISO in the software industry. It may be followed by specific training in the areas of implementation strategies, writing the documentation, and auditing training. In addition, it may be followed by the TickIT program, which was explained in Chapter 8.

I. An Introduction to ISO 9000
   A. What is it?
   B. What are the driving forces?
   C. TickIT
   D. RAB initiatives in the United States
   E. U.S. companies registered
   F. Benefits to be gained
   G. Control of the development process and life cycle
   H. Improved development cycle time and quality
   I. Reductions in software errors
   J. Reduced software development and maintenance costs
   K. Who runs the programs?
II. Salient points of ISO 9001
   A. Highlights of the 20 clauses
   B. Relationship to software
III. What is ISO 9000-3?

    A.   Where did it come from and how was it developed?

    B.   What is its purpose?

    C.   Is it a standard or a guideline?

    D.   Why is it necessary indeed?

IV.   Structure of ISO 9000-3

    A.   How is it organized and why?

    B.   Definitions and terminology

    C.   Cross-reference between ISO 9001 and ISO 9000-3

V.   Explaining ISO 9000-3

    A.   A departure from the usual clause-by-clause presentation

    B.   How the guidance of ISO 9000-3 can be used to satisfy the requirements of ISO 9001

VI.   Summary of ISO 9000-3

    A.   A depiction of a 9001-conforming quality system for software development

    B.   Implications to the industry

VII.   How to get registered

    A.   Steps to registration

    B.   Additional help available

    C.   What options are available

    D.   What an ISO auditor will expect to see for software quality

GMP COMPLIANCE AND ISO 9000/Q90 REGISTRATION

The intent of this executive training course is to familiarize those who develop (and/or design) and manufacture medical devices in compliance with the FDA's Good Manufacturing Practices (GMPs) and now need to understand the world of ISO 9000. It may be followed by specific training in the areas of implementation strategies, writing the documentation, and auditing training.

    I.   Introduction

   II.   Course objectives

  III.   GMP and ISO 9000: an overview

  A. What is ISO 9000?

  B. What is GMP?

  C. Why do you need it?

  D. Who runs the programs?

IV. GMP/ISO compliance matrix

V. Bridging the gap

VI. The steps to registration

  A. Organize the effort

    1. Obtain management commitment

    2. Organize management and organizational training awareness

    3. Appoint a management representative

    4. Define project structure

    5. Prepare implementation groups

  B. Establish a management steering committee

VII. Document the system

  A. Define the documentation needed

  B. Identify the driving forces for completing the documentation

  C. Identify the required documentation: formal and informal

  D. Define the most appropriate hierarchy for your documentation

    1. Quality manual

    2. Quality procedures

    3. Work instructions

    4. Forms and tags

VIII. Consolidation of people, processes, and documentation

  A. Understand the ISO standard in relation to the philosophy of "continual improvement"

  B. Train in audit system effectiveness

  C. System to report to steering committee

  D. Appropriate ALL necessary resources for training

  E. Conduct a preassessment

IX. Registration strategy

  A. Initial contacts

  B. Initial response

       C. Lead time
       D. Preassessment services
       E. Costs
       F. Develop a relationship (your organization with the registrar)
       G. Be prepared for questions and more questions
  X. Register your company
       A. Define the scope of registration
       B. Prepare for the site audit
       C. The registration audit
       D. Audit management
  XI. Definitions

TRAINING FOR UNDERSTANDING AND IMPLEMENTING THE ISO 9000 SERIES

This 2- or 3-day training course provides a thorough understanding for the ISO 9000 theory, history, development, structure, and implementation methodology for the certification to the appropriate standard (ISO 9001, 9002, or 9003). It also provides the participants with an understanding of what to expect in the registration process—from preparation, documentation, and auditing to certification and recertification. The length depends on the number of workshops the training will accommodate. Otherwise, the material is exactly the same.

  I. Overview of ISO 9000 series
       A. History of EC
       B. Development of European standards
       C. European Community institutions
       D. Directives
       E. Related standards and guidelines
       F. Status of directives
       G. ISO 9000 modules
       H. Types of products
       I. Agreement groups
       J. Sectoral committees
       K. Cross-reference

        L.   Conformity assessment
II.   Overview of accreditation
        A.   Need
        B.   System
        C.   Registration in the United States
        D.   Process for registration
            1.   Factors
            2.   Routes
            3.   Getting registered
        E.   Steps to quality system registration
            1.   Budget work sheet
            2.   RAB auditors
            3.   Specific industry requirements
        F.   Stages of quality evolution
            1.   Quality standards and awards
            2.   DuPont model
            3.   MBNQA
III.   ISO quality assurance standards overview
        A.   Definitions
        B.   Requirements
        C.   System elements
        D.   Comparison
        E.   Quality loop
        F.   Purpose
        G.   Advantages
        H.   Limitations
        I.   Shortcomings
        J.   Benefits
        K.   Elements of certification
            1.   Supplier approval
            2.   Quality manual
            3.   Documents
            4.   Reviews
        L.   Process for developing compliance
IV.   ISO 9000 standard overview
        A.   Relationship and concepts
        B.   Types of standards

    C. Use of standards
    D. Selection of factors
    E. Effect of choice
    F. Demonstration of system
    G. Precontract assessment
V. Quality system models
    A. ISO 9001
        1. Definition
        2. Detailed explanation by each individual clause
    B. ISO 9004
        1. Definition
        2. Guide to quality
VI. Audits overview
    A. Types
    B. Definition
    C. Scope
    D. Phases
    E. Preparation
    F. Performance
VII. Implementation strategy
    A. Establish a management steering committee
    B. Appoint an ISO facilitator (project manager)
    C. Define the documentation needed for your organization
    D. Train your force in the ISO standards
    E. Prepare for registration

ISO 9000 DOCUMENTATION TRAINING

This 2-day training course provides the minimum requirements for an in-depth review of ISO 9000 requirements and methods for documentation.

I. Overview of ISO 9000 and ISO 9004
    A. What is ISO?
    B. What is ISO 9000 and ISO 9004?
    C. Why do we need ISO?

      D.   Who runs the program?
      E.   How do we get registered?
      F.   Is there a connection between documentation and registration? If so, what is it?

II.   The importance of documentation
      A.   What drives the documentation process?
      B.   Benefits of proper documentation
      C.   What is appropriate and applicable documentation for the ISO?

III.   ISO 9000 requirements
      A.   Flexibility of requirements
      B.   The ISO 9004 factor in the documentation process
      C.   Formal versus informal documentation

IV.   Format for documentation
      A.   Structuring the documents
          1.   Hierarchy
          2.   Parsing
      B.   Terminology
      C.   Logical conventions and structure of writing
      D.   Organization's conventions and structure of writing
      E.   Cross-referencing

V.   Contents of appropriate documentation
      A.   Company mission and policy statements
      B.   Quality system documentation
          1.   The quality manual
          2.   The procedures
          3.   The instructions
          4.   Forms
          5.   Quality records
          6.   Quality plans

VI.   How to write the documentation
      A.   Company mission and policy statements
      B.   Quality system documentation
          1.   The quality manual
          2.   The procedures
          3.   The instructions
          4.   Forms

5. Quality records
6. Quality plans
C. The quality standards and procedures development process

VII. Relationship of project life cycle documents and design control
VIII. Document control
A. The role of ISO 9004
B. Variety of control method
C. Approval authority
D. Revision history
E. Handwritten changes
F. Obsolete documents
IX. Document production issues
A. Effective documentation
1. Keep it simple
2. Know the audience
B. Principles of good technical writing
1. Form and format
2. Structure
3. Pagination
4. Appearance
5. Use of flowcharts, pictures, diagrams, figures, and tables
C. Editing, reviews, and proofreading techniques
D. Appropriate reading level

ISO 9000 INTERNAL AUDITOR TRAINING

This 2-day training course explains why auditing is an essential component of every quality assurance system and how to carry out audits to comply with ISO 9000 and ISO 10011 requirements. It is diverse enough to be applied across all industries. Any training in auditing should provide role playing for the auditors as well as exercises to enhance audit and communication skills.

I. Quality and quality assurance
A. Quality defined

      B.   Quality assured

      C.   Quality controlled

  II. Quality audits

      A.   What is an audit?

      B.   Why audit?

      C.   Overview of audits

 III. Auditing to the standard

      A.   ISO 9000 theory and structure

      B.   Documentation review

      C.   The ISO 9001 explained

 IV. Audit process

      A.   General information

      B.   Types of audits

      C.   Characteristics

      D.   Who is involved

  V. Audit administration

      A.   Lead auditor training qualifications

      B.   Auditor training qualifications

      C.   The use of lead auditor versus auditor

      D.   Maintenance of qualification records

      E.   Appropriate audit records

          1.   System analysis

          2.   Checklists

          3.   Tools

          4.   Working papers

      F.   Evaluating the evaluators

      G.   ISO 10011 requirements

      H.   Sampling

      I.   Planning for the appropriate resources

 VI. The auditor

      A.   A position of trust

      B.   Auditor qualifications

      C.   Auditor traits, knowledge, and aptitude

      D.   Ethics

VII. The audit

      A.   Types

          1.   Scope

          2.   Purpose
     B. Audit preparation
          1.   Define the scope of the audit
          2.   Define the resources of the audit
          3.   Define the time of the audit
          4.   Schedule the audit
          5.   Prepare a checklist
          6.   Prepare an audit plan
          7.   Send (if appropriate) the notification letter

VIII. The functional audit
     A. Preparation
     B. The opening meeting
     C. The actual audit
     D. Documenting nonconformances
     E. Team meetings
     F. Exit meetings
     G. Communicate results of the audit
     H. Inform of the "certification" status

  IX. Audit closure
     A. Reports and follow-up
     B. Act on corrective action requests

   X. Workshops
     A. Several activities to enhance the understanding of the actual standards
     B. Audit role play
     C. A lengthy exercise on a "make-believe" audit

ISO 9000 LEAD AUDITOR TRAINING

This 5-day training course prepares the participant to conduct and be in charge—if the participant has all the other requirements—of an ISO 9000 audit. It is very extensive and concentrated training requiring anywhere from 12 to 15 hours daily. Teamwork is expected and after-hours work is mandatory.

Day 1: Welcome and introductions

       Quality and QA

What is ISO 9000?
Who runs the program?
Definitions
Quality management systems ISO 9000
   Analysis of the individual clauses
   Requirements of the ISO
      Documentation
      Quality manual
      Procedures
      Instructions
      Forms and tags
Certification and assessment Workshops
   Extensive exercises on the entire set of the standards
   Read a quality manual for discussion next day

Day 2: Questions and concerns about the company represented in the quality manual

The assessment/audit system
   Types
   A perspective on first-, second-, and third-party assessments
      Scope
      Objective
   Quality assurance manuals
   Plan the audit
   Prepare a checklist
   Evaluate the checklist
   Prepare a notification letter
   Prepare an audit plan
   Prepare an audit schedule
Workshops
   Extensive workshops on preparing a checklist, notification letter, audit plan, and an audit schedule
   Prepare a checklist based on specific clauses of the standard from the quality manual that you have been given

Day 3: Presentation and discussion of the checklist and general preparation of the audit
Feedback
Opening meeting

    Carrying out the audit

    Nonconformances

        Found

        Recorded

    Workshops

        Extensive workshops on the mechanics of the audit

        Writing and presenting noncompliances

        Prepare action items from the quality manual and case studies given to you

Day 4: Present and discuss the noncompliances as well as the action items

    Feedback

    Closing meeting

    Reporting the audit

    Corrective action

    Follow-up and surveillance

    Workshops:

        Role playing

        Extensive writing practice on reporting noncompliance items, corrective actions, and follow-up requirements

        Review material for next day's examination

Day 5: Present and discuss the writing exercises

    Feedback

    Summary

    Review of the training

    Examination

Appendix D

General Information and Publications

1. STANDARDS CODE AND INFORMATION PROGRAM (SCI): NATIONAL INSTITUTE OF STANDARDS AND TECHNOLOGY

The ABC's of Standards-Related Activities in the United States (NBSIR 87-3576)

This report is an introduction to voluntary standardization, product certification, and laboratory accreditation for readers not fully familiar with these topics. It stresses some of the important aspects of these fields, furnishes the reader with both historical and current information on these topics, describes the importance and impact of the development and use of standards, and serves as background for using available documents and services.

Order as PB 87-224309 from NTIS.

The ABC's of Certification Activities in the United States (NBSIR 88-3821)
This report, a sequel to NBSIR 87-3576, provides an introduction to certification for readers not entirely familiar with this topic. It highlights some of the important aspects of this field, furnishes the reader with information necessary to make informed purchases, and serves as background for using available documents and services.
Order as PB 88-239793 from NTIS.

Laboratory Accreditation in the United States (NISTIR 4576)
This report, a sequel to NBSIR 87-3576 and NBSIR 88-3821, is designed to provide information on laboratory accreditation to readers who are new to this field. It discusses some of the significant facets of this topic, provides information necessary to make informed decisions on the selection and use of laboratories, and serves as background for using other available documents and services.
Order as PB 91-194495 from NTIS.

Questions and Answers on Quality, the ISO 9000 Standard Series, Quality System Registration, and Related Issues (NISTIR 4721)
This report provides information on the development, content, and application of the ISO 9000 standards to readers who are unfamiliar with these aspects of the standards. It attempts to answer some of the most commonly asked questions on quality; quality systems; the content, application, and revision of the ISO 9000 standards; quality system approval/registration; European Community requirements for quality system approval/registration; and sources for additional help.
Copies not available from SCI. Order as PB 92-126465 from NTIS.

Directory of International and Regional Organizations Conducting Standards-Related Activities (NIST SP 767)
This directory contains information on 338 international and regional organizations that conduct standardization, certification, laboratory accreditation, or other standards-related activities. It describes their work in these areas, as well as the scope of each organization, national affiliations of members, U.S. participants, restrictions on membership, and the availability of any standards in English.
Copies not available from SCI. Order as PB 89-221147 from NTIS or order as Cat. #sp767 from Global Engineering Documents.

Directory of European Regional Standards-Related Organizations (NIST SP 795)

This directory identifies more than 150 European regional organizations—both governmental and private—that engage in standards development, certification, laboratory accreditation and other standards-related activities, such as quality assurance. Entries describe the type and purpose of each organization, acronyms, national affiliations of members, the nature of the standards-related activity, and other related information.

Copies not available from SCI. Order as PB 91-107599 from NTIS or order as Cat. #0258-3 from Global Engineering Documents.

Standards Activities of Organizations in the United States (NIST SP 806)

This directory identifies and describes activities of over 750 U.S. public and private-sector organizations that develop, publish, and revise standards; participate in this process; or identify standards and make them available through information centers or distribution channels. NIST SP 806, a revision of NBS SP 681, covers activities related to both mandatory and voluntary U.S. standards. SP 806 also contains a subject index and related listings that cover acronyms and initials, defunct bodies, and organizations with name changes.

Copies not available from SCI. Order as PB 91-177774 from NTIS or order as Cat. #Sp806 from Global Engineering Documents.

Directory of Private Sector Product Certification Programs (NIST SP 774)

This directory presents information from 132 private-sector organizations in the United States that engage in product certification activities. Entries describe the type and purpose of each organization, the nature of the activity, product certified, standards used, certification requirements, availability and cost of services, and other relevant details.

Copies not available from SCI. Order as PB 90-161712 from NTIS.

Directory of Federal Government Certification Program (NBS SP 739)

This directory presents information on U.S. government certification programs for products and services. Entries describe the scope and nature of each certification program, testing and inspection practices, standards used, methods of identification and enforcement, reciprocal recognition or acceptance of certification, and other relevant details. Copies not available from SCI. Order as PB 88-201512 from NTIS.

Directory of Federal Government Laboratory Accreditation/Designation Programs (NIST SP 808)

This directory provides updated information on 31 federal government laboratory accreditation and similar-type programs conducted by the federal government. These programs, which include some type of assessment regarding laboratory capability, designate sets of laboratories or other entities to conduct testing to assist federal agencies in carrying out their responsibilities. The directory also lists 13 other federal agency programs of possible interest, including programs involving very limited laboratory assessment and programs still under development. Copies not available from SCI. Order as PB 91-167379 from NTIS.

Directory of State and Local Government Laboratory Accreditation/ Designation Programs (NIST SP 815)

This directory provides updated information on 21 state and 11 local government laboratory accreditation and similar-type programs. These programs, which include some type of assessment regarding laboratory capability, designate private-sector laboratories or other entities to conduct testing to assist state and local government agencies in carrying out their responsibilities. Entries describe the scope and nature of each program, laboratory assessment criteria and procedures used in the program, products and fields of testing covered, program authority, and other relevant details. Copies not available from SCI. Order as PB 92-108968 from NTIS.

Directory of Professional/Trade Organization Laboratory Organization Laboratory Accreditation/Designation Programs (NIST SP 831)

This directory is a guide to laboratory accreditation and similar types of programs conducted by professional and trade organizations. These programs accredit or designate laboratories or other entities to assist private-sector professional societies, trade associations, related certification bodies, their membership, as well as government agencies, in carrying out their responsibilities. This accreditation or designation is based on an assessment of the capability of the laboratory to conduct the testing. However, the nature of the assessment varies considerably by organization and program. Order as SN 003-003-03244-5 from GPO.

Barriers Encountered by U.S. Exporters of Telecommunications Equipment (NBSIR 87-3641)
This report addresses the perceived institution of unreasonable technical trade barriers by major European trading partners to the export of telecom products and systems by U.S. companies. The GATT technical office, which has responsibilities to assist U.S. exporters to take advantage of trade opportunities, informally contacted, over a period of 6 months, telecom companies and agencies to assess the extent of unreasonableness in foreign national standards, regulations, testing and certification requirements, and accreditation procedures.
Copies not available from SCI. Order as PB 88-153630 from NTIS.

A Review of U.S. Participation in International Standards Activities (NBSIR 88-3698)
This report describes the role of international standards, their increasingly significant importance in world trade, and the extent of past and current U.S. participation in the two major international standardization bodies—ISO and IEC. The degree of U.S. participation covers the 20-year period 1966–1986. A course analysis of data indicates some correlation between U.S. participation and recent export performance for several major product categories.
Copies not available from SCI. Order as PB 88-164165 from NTIS.

An Update of U.S. Participation in International Standards Activities (NISTIR 89-4124)
This report presents updated information on the current level of U.S. participation in ISO and IEC. (Reference: NBSIR 88-3698.)
Copies not available from SCI. Order as PB 89-228282/AS from NTIS.

A Summary of the New European Community Approach to Standards Development (NBSIR 88-3793-1)
This paper summarizes European Community plans to aggressively pursue its goal of achieving an "internal market" by 1992 and the standards-related implication of such a program on U.S. exporters.
Order as PB 88-229489/AS from NTIS.

Trade Implications of Processes and Production Methods (PPMs) (NISTIR 90-4265)
This report discusses processes and production methods (PPMs) and their relationship to trade, the GATT Agreement on Technical Barriers

to Trade, and traditional product standards used in international commerce. The report provides background information on PPMs, a suggested definition, and the possible extension of their application from the agricultural sector to industrial products.
Order as PB 90-205485 from NTIS.

2. DOCUMENTS AVAILABLE UPON REQUEST FROM SCI

tbt news
This newsletter provides information on government programs and available services established in support of the GATT Agreement on Technical Barriers to Trade (Standards Code). *tbt news* reports on the latest notifications of proposed foreign regulations, bilateral consultations with major U.S. trade partners, programs of interest to U.S. exporters, and availability of standards and certification information. Subscription is free upon request.

Technical Barriers to Trade
This booklet explains the basic rules of the international Agreement of Technical Barriers to Trade negotiated during the Tokyo Round of the Multilateral Trade Negotiations (MTN) and describes Title IV of the U.S. Trade Agreements Act of 1979, which implements the United States' obligations under the agreement. The agreement, popularly known as the Standards Code, was designed to eliminate the use of standards and certification systems as barriers to trade. The booklet describes the functions of the Departments of Commerce and Agriculture, the Office of the U.S. Trade Representative, and the State Department in carrying out the United State's responsibilities.

GATT Standards Code Activities
This brochure gives a brief description of NIST's activities in support of the Standards Code. These activities include operating the U.S. GATT inquiry point for information on standards and certification systems, notifying the GATT Secretariat of proposed U.S. regulations, assisting U.S. industry with trade-related standards problems, responding to inquiries on foreign and U.S. proposed regulations, and preparing reports on the Standard Code.

GATT Standards Code Activities of the National Institute of Standards and Technology

This annual report describes the GATT Standards Code activities conducted by the Standards Code and Information Program for each calendar year. NIST responsibilities include operating the GATT inquiry point, notifying the GATT Secretariat of proposed U.S. federal government regulations that may affect trade, assisting U.S. industry with standards-related trade problems, and responding to inquiries about proposed foreign and U.S. regulations.

Free handout material on office activities and standards-related information, such as: government sources of specifications and standards, foreign standards bodies, U.S. standards organizations, and a brochure on the National Center for Standards and Certification Information (NCSCI).

In addition to general inquiry services, the following assistance is also available:

EC Hotline

This hotline reports on draft standards of the European Committee on Standardization (CEN), the European Committee for Electrotechnical Standardization (CENELEC), and the European Telecommunications Standards Institute (ETSI). It also provides information on selected EC directives. The recorded message is updated weekly and gives the product, document number, and closing date for comments. The hotline number is (301) 921-4164 (not toll free).

GATT Hotline

A telephone hotline provides current information received from the GATT Secretariat in Geneva, Switzerland, on proposed foreign regulations that may significantly affect trade. The recorded message is updated weekly and gives the product, country, closing date for comments (if any), and Technical Barriers to Trade (TBT) notification number. The hotline number is (301) 975-4041 (not toll free).

NCSI provides assistance to U.S. and foreign exporters in obtaining current standards, regulations, and certification information for the manufacture of products. To aid foreign exporters, NCSCI also provides directory information of state offices prepared to respond to queries concerning conditions to be met by goods for sale in their state.

3. SOURCES FOR OBTAINING THESE PUBLICATIONS

National Technical Information Services (NTIS)
5285 Port Royal Road
Springfield, VA 22161, USA
Telephone: (703) 487-4650
Orders only: (800) 553-6847
Fax: (703) 321-8547

Superintendent of Documents
U.S. Government Printing Office (GPO)
Washington, DC 20402, USA
Telephone: (202) 783-3238
Fax: (202) 512-2250

Global Professional Publications
15 Inverness Way East, P.O. Box 1154
Englewood, CO 80150-1154, USA
Telephone: (800) 854-7179
Local phone: (303) 792-2181
Fax: (303) 792-2192

When requesting publication information from SCI, send a self-addressed mailing label to:

Standards Code and Information Program (SCI)
National Institute of Standards and Technology
Administration Building, Room A629
Gaithersburg, MD 20899, USA

For assistance in obtaining information on current U.S. and foreign standards, regulations, and certification information, contact:

The National Center for Standards and Certification Information (NCSCI)
National Institute of Standards and Technology
TRF Building, Room A163
Gaithersburg, MD 20899, USA
Telephone: (301) 975-4040

4. SOURCES FOR ORDERING STANDARDS

(Copies can be obtained from the respective standards-issuing organization and/or these sources.)

Organization	Information provided
American National Standards Institute (ANSI) 11 West 42nd Street, 13th Floor New York, NY 10036, USA Telephone: (212) 642-4900 Fax: (212) 398-0023 Orders only: (212) 302-1286 Telex: 42 42 96 ANSI UI	ANSI-approved industry standards International and foreign standards Select draft CEN/CENELEC standards; draft ISO standards
Global Professional Publications 15 Inverness Way East, P.O. Box 1154 Englewood, CO 80150-1154, USA Specifications Telephone: (800) 854-7179 Local phone: (303) 792-2181 Fax: (303) 792-2192	Industry standards Federal standards and specifications Military standards International and foreign standards
National Standards Association (NSA) 1200 Quince Orchard Boulevard Gaithersburg, MD 20878, USA Documents Telephone: (800) 638-8094 Local phone: (301) 590-2300 Fax: (301) 990-8378 Telex: 44 6194 NATSTA GAIT	Industry standards Federal and military standards, specifications, and related data NATO standards Aerospace standards
General Services Administration (GSA) Federal Supply Service Bureau Specifications Branch 490 East L'Enfant Plaza, SW Suite 8100 Washington, DC 20407, USA Telephone: (202) 755-0325 or 0326 Fax: (202) 205-3720	Federal standards and specifications

Naval Publications and Forms Center
ATTN: NPODS
5801 Tabor Avenue
Philadelphia, PA 19120-5099, USA
Inquiries (not for placing orders)
Telephone: (215) 697-2667
Fax: (215) 697-5914
	Department of Defense (DOD)
	  adopted documents
	Naval publications
	Military manuals and other related
	  forms

Standardization Document Order
  Desk Naval Publications Printing
  Service
700 Robbins Avenue, Building 4,
  Section D
Philadelphia, PA 19111-5094, USA
Telephone: (215) 697-2179
Fax: (215) 697-2978
	Military standards, specifications,
	  handbooks
	Federal standards and
	  specifications

Document Center 1504 Industrial Way, Unit 9 Belmont, CA 94002, USA Specifications Telephone: (415) 591-7600 Fax: (415) 591-7617	Industry standards Federal standards and specifications Military standards International and foreign standards

Information Handling Services (IHS)
  (for IHS subscribers only)
P.O. Box 1154
Iverness Way East
Specifications
Englewood, CO 80150, USA
(CEN/CENELEC)
Telephone: (800) 241-7824
Local phone: (303) 790-0600
Fax: (303) 799-4097
Telex: 4322083 IHS UI
	International and foreign standards
	Industry standards
	Federal standards and
	  specifications
	Military standards
	Select European standards

Standards Sales Group (SSG)
9420 Reseda Boulevard, Suite 800
Northridge, California 91324, USA
Information and quotes:
Telephone: (818) 368-2786
Orders only: (800) 755-2780
Fax: (818) 360-3804
	International and foreign standards
	Publications and other reference
	  materials
	Translations service
	U.S./Foreign general regulatory
	  compliance
	Information

5. ISO SOME RELEVANT DOD STANDARDS

MIL-Q-9858A—Quality program requirements
MIL-I-45208A—Inspection system requirements

6. STANDARDS/GUIDES

ISO/IEC Guide 2, General Terms and their Definitions Concerning Standardization and Related Activities

ISO/IEC Guide 7, Requirements for Standards Suitable for Product Certification

ISO/IEC Guide 16, Code of Principles on Third-Party Certification and Related Standards

ISO/IEC Guide 22, Information on Manufacturer's Declaration of Conformity with Standards or Other Technical Specifications

ISO/IEC Guide 23, Methods for Indicating Conformity with Standards for Third-Party Certification Systems

ISO/IEC Guide 25, General Requirements for the Competence of Calibration and Testing Laboratories

ISO/IEC Guide 27, Guidelines for Corrective Action to be Taken by a Certification Body in the Event of Misuse of Its Mark of Conformity

ISO/IEC Guide 28, General Rules for a Model Third-Party Certification System for Products

ISO/IEC Guide 39, General Requirements for the Acceptance of Inspection Bodies

ISO/IEC Guide 40, General Requirements for the Acceptance of Certification Bodies

ISO/IEC Guide 43, Development and Operation of Laboratory Proficiency Testing

ISO/IEC Guide 44, General Rules for ISO or IEC International Third-Party Certification Schemes for Products

ISO/IEC Guide 45, Guidelines for the Presentation of Test Results

ISO/IEC Guide 46, An Approach to the Review by a Certification Body of Its Own Internal Quality System

ISO/IEC Guide 48, Guidelines for Third-Party Assessment and Registration of a Supplier's Quality System

ISO/IEC Guide 53, An Approach to the Utilization of a Supplier's Quality System in Third-Party Product Certification

Draft ISO/IEC Guide 58, Calibration and Testing Laboratory Accreditation System

General Requirements for Operation and Recognition (Revision of ISO/IEC Guides 25, 54, and 55)

ISO 8402, Quality—Terminology

ISO 9000 (ANSI/ASQC Q 91/EN 29001), Quality Systems—Model for Quality Assurance in Design/Development, Production, Installation, and Servicing

ISO 9002 (ANSI/ASQC Q 92/EN 29002), Quality Systems—Model for Quality Assurance in Production and Installation

ISO 9003 (ANSI/ASQC Q 93/EN 29003), Quality Systems—Model for Quality Assurance in Final Inspection and Test

ISO 9004 (ANSI/ASQC Q 94/EN 29004), Quality Management and Quality System Elements—Guidelines

ISO 10011 Part 1, Guidelines for Auditing Quality Systems—Auditing

ISO 10011 Part 2, Guidelines for Auditing Quality Systems—Qualification Criteria for Auditors

ISO 10011 Part 3, Guidelines for Auditing Quality Systems—Management of Audit Programmes

ISO 10012-1, Quality Assurance Requirements for Measuring Equipment—Part 1: Metrological Confirmation System for Measuring Equipment

7. CEN/CENELEC—EN 45000 AND 46000 STANDARDS

EN 45001, General Criteria for the Operation of Testing Laboratories

EN 45002, General Criteria for the Assessment of Testing Laboratories

EN 45003, General Criteria for Laboratory Accreditation Bodies

EN 45011, General Criteria for Certification Bodies Operating Product Certification

EN 45012, General Criteria for Certification Bodies Operating Quality System Certification

EN 45014, General Criteria for Supplier's Declaration of Conformity

EN 45020 (ISO/IEC Guide 2), General Terms and Their Definitions Concerning Standardization and Related Activities

prEN 46001, Specific Requirements for the Application of EN 29001 for
  Medical Devices

8. SOURCES FOR ADDITIONAL INFORMATION ON NIST-RELATED ACTIVITIES AND ORGANIZATIONS/ DOCUMENTS REFERENCED IN THE TEXT

NIST's National Voluntary Laboratory Accreditation Program (NVLAP)
  NVLAP/NIST
  Building 411, Room A162
  Gaithersburg, MD 20899
  Phone: (301) 975-4042
  Fax: (301) 926-2884

The Malcolm Baldrige National Quality Award Program
  Office of Quality Programs/NIST
  Building 101, Room A537
  Gaithersburg, MD 20899
  Phone: (301) 975-3771

NIST's Calibration Program
  Calibration Program/NIST
  Building 411, Room A104
  Gaithersburg, MD 20899-0001
  Phone: (301) 975-2002
  Fax: (301) 926-2884

NIST's Standard Reference Materials Program
  Standard Reference Materials Program/NIST
  Building 202, Room 204
  Gaithersburg, MD 20899-0001
  Phone: (301) 975-6776
  Fax: (301) 948-3730

NIST's Standard Reference Data Program
  Standard Reference Data Program/NIST
  Building 221, Room A320
  Gaithersburg, MD 20899-0001

Phone: (301) 975-2208
Fax: (301) 926-0416

The American Society for Quality Control's (ASQC) standards, pub-
 lications, activities, and services; ISO TC 176's activities
American Society for Quality Control (ASQC)
611 East Wisconsin Ave.
P.O. Box 3005, Milwaukee, WI 53202
Phone: (414) 272-8575
Fax: (414) 765-8661

The Registrar Accreditation Board's (RAB) program
 Registrar Accreditation Board (RAB)
 611 East Wisconsin Ave.
 P.O. Box 3005, Milwaukee, WI 53202
 Phone: (414) 272-8575
 Fax: (414) 765-8661

ANSI's activities or to purchase copies of ISO draft/final standards,
 other documents, magazines, and newsletter, and/or copies of
 European standards (ENs)
 The American National Standards Institute (ANSI)
 11 West 42nd Street, 13th Floor, New York, NY 10036
 Phone: (212) 642-4900
 Fax: (212) 302-1286

CEEM's Registered Company Directory, *Quality Systems Update*
 newsletter, and other publications
 CEEM
 10521 Braddock Road
 Fairfax, VA 22032
 Phone: (800) 745-5565 or (703) 250-5900
 Fax: (703) 250-5313

The Aerospace Industries Association's (AIA) activities
 Aerospace Industries Association
 1250 Eye Street, NW
 Washington, DC 20005
 Phone: (202) 371-8400

The Netherlands' RvC program
  Raad voor de Certificatie
  Stationseg 13F, 3972 KA Driebergn
  Phone: +31 34 381 26 04
  Fax: +31 34 381 85 54

The British Institute of Quality Assurance's (IQA) quality system asses-
      sor registration program
  The Secretary to the Board
  National Registration Scheme for Assessors of Quality Systems
  The Institute of Quality Assurance
  10 Grosvenor Gardens, London, U.K. SWIW ODQ
  Phone: 44-71-730-7154

The Standards Council of Canada (SCC) Program
  Standards Council of Canada (SCC)
  45 O'Connor Street, Suite 1200
  Ottawa, Ontario K1P 6N7 Canada
  Phone: (613) 238-3222
  Fax: (613) 995-4564

The European Organization for Testing and Certification
  The EOTC
  Rue Stassart 33, 2nd Floor
  B-1050 Brussels, Belgium
  Phone: +32 2 519 6969
  Fax: +32 2 519 69 17/19

The International Electrotechnical Commission's (IEC) Quality Assess-
      ment System for Electronic Components (IECO System)
  Electronic Components Certification Board (ECCB)
  Electronic Industries Association (EIA)
  2001 Pennsylvania Ave, NW Washington, DC 20006
  Phone: (202) 457-4967

Automotive Industry Action Group (AIAG)
  26200 Lahser Rd., Suite 200
  Southfield, MI 48034
  Phone: (810) 358-3570
  Fax: (810) 799-4220

9. MANUFACTURING TECHNOLOGY CENTERS (MTCs)

These seven regional centers were established by NIST to serve as resource facilities to help manufacturers improve their competitive position through the application of manufacturing technology.

NORTHEAST MTC
Mr. Mark S. Tebbano, Director
Northeast MTC
NY State Science & Tech
  Foundation
99 Washington Ave.
Albany, NY 12210
Phone: (518) 473-9746

SOUTHEAST MTC
Mr. James Bishop, Director
Southeast MTC
P.O. Box 1149
Columbia, SC 29202
Phone: (803) 737-0410

GREAT LAKES MTC
Dr. George Sutherland, Director
Great Lakes MTC
2415 Woodland Ave.
Cleveland, OH 44115
Phone: (216) 987-3201

CALIFORNIA MTC
Dr. John Chernesky, Director
California Community College
1107 Ninth Street
Sacramento, CA 95814

MIDWEST MTC
Dr. George H. Kuper, Acting
  Director
Industrial Technology Institute
P.O. Box 1485
2901 Hubbard Road
Ann Arbor, MI 48109
Phone: (313) 769-4710

MID-AMERICA MTC
Mr. Paul Clay, President
Mid-America MTC
10561 Barkley, Suite 602
Overland Park, KS 66212
Phone: (913) 649-4333

MINNESOTA TECHNOLOGY,
  INC.
Ms. Jane Pounds, Director
111 Third Ave.
Minneapolis, MN 55401
Phone: (612) 338-7722

10. TRADE ADJUSTMENT ASSISTANCE CENTERS (TAACs)

The Department of Commerce's Economic Development Administration (EDA) funds 12 regional Trade Adjustment Assistance Centers to help ailing companies.

NEW ENGLAND TAAC
Richard McLaughlin, Director
New England TAAC
120 Boylston Street
Boston, MA 02116
Phone: (617) 542-2395
Fax: (617) 542-8457
(CT, RI, VT, NH, MA, ME)

NEW JERSEY TAAC
John Walsh, Acting Director
NJ Economic Development
  Authority
Capital Place One-CN 990
200 South Warren Street
Trenton, NJ 08625
Phone: (609) 292-0360
Fax: (609) 292-0368
(NJ)

NEW YORK TAAC
John Lacey, Director
New York TAAC
117 Hawley Street, Suite 102
Binghamton, NY 13901
Phone: (607) 771-0875
Fax: (607) 724-2404
(NY)

SOUTHEASTERN TAAC
Charles Estes, Acting Director
Georgia Institute of Technology
Research Institute
151 6th St.
O'Keefe Building, Room 224
Atlanta, GA 30332
Phone: (404) 894-3858, 6789,
  6106
Fax: (404) 853-9172
(AL, TN, KY, MS, GA, NC, SC,
  FL)

SOUTHWEST TAAC
Robert Velasquez, Director
301 South Frio St., Suite 225
San Antonio, TX 78207
Phone: (210) 220-1240
Fax: (210) 220-1241
(TX, LA, OK)

MID-AMERICA TAAC
Paul Schmid, Director
University of Missouri at
  Columbia
University Place, Suite 1700
Columbia, MO 65211
Phone: (314) 882-6162
Fax: (314) 882-6156
(MO, KS, AR)

MID-ATLANTIC TAAC
William Gates, Director
Mid-Atlantic TAAC
486 Norristown Road
Suite 130
Blue Bell, PA 19422
Phone: (215) 825-7819
Fax: (215) 825-7834
(PA, DE, MD, VA, WV, DC)

MIDWEST TAAC
Howard Yefsky, Director
Applied Strategies Int'l
150 N. Wacker Dr. Suite 2240
Chicago, IL 60606
Phone: (312) 368-4600
Fax: (312) 368-9043
(IL, MN, IA, WI)

NORTHWEST TAAC
Ronald Horst, Director
Bank of California Center
900 4th Avenue, Suite 2430
Seattle, WA 98164
Phone: (206) 622-2730
Fax: (206) 622-1105
(AK, ID, MT, OR, WA)

GREAT LAKES TAAC
Margaret Creger, Director
University of Michigan
School of Business
    Administration
506 Easy Liberty Street
Ann Arbor, MI 48104-2210
Phone: (313) 998-6213
Fax: (313) 998-6224
(MI, OH, IN)

ROCKY MOUNTAIN TAAC
Robert Stansbury, Director
Rocky Mountain TAAC
5353 Manhattan Cir., Suite 200
Boulder, CO 80303
Phone: (303) 499-8222
Fax: (303) 499-8298
(CO, UT, NE, SD, ND, WY,
NM)

WESTERN TAAC
Daniel Jimenez, Director
USC-WTAAC
3716 S. Hope Street, Suite 200
Los Angeles, CA 90007
Phone: (213) 743-8427
Fax: (213) 746-9043
(AZ, CA, NV, HI)

Selected Bibliography

Anon. (November 1992). "A strategy for standards and quality." *Quality at Work Newsletter*. Yarsley Quality Assured Firms Ltd.

Anon. (April 1993). "How to implement ISO 9000." *Continuous Improvement*. National ISO 9000 Support Group, Alto, MI.

Anon. (July 1993). "ISO 9000 and inspection." *Continuous Improvement*. National ISO 9000 Support Group, Alto, MI.

Anon. (July/August/September 1993). "Environmental leadership." *Leaders*.

Anon. (August 1993). "More questions about ISO 9000." *Continuous Improvement*. National ISO 9000 Support Group, Alto, MI.

Anon. (August 1993). "The environmental standard—BS7750." *Continuous Improvement*. National ISO 9000 Support Group, Alto, MI.

Anon. (September 1993). "The environmental standard—BS7750." *Continuous Improvement*. National ISO 9000 Support Group, Alto, MI.

Anon. (October 1993). "How to prepare for ISO 9000." *Continuous Improvement*. National ISO 9000 Support Group, Alto, MI.

Anon. (November 1993). "Health care TC forming." *ISO 9000 Journal*, Vol. J-3.

Anon. (December 1993). "Tips on streamlining your ISO 9000 certification process." *Continuous Improvement*. National ISO 9000 Support Group, Alto, MI.

Anon. (March 1994). "Control software: Function point analysis and controls software." *ESD Technology*. Engineering Society of Detroit, Detroit, MI.

Ali, N. S. (April 1994). "Developing 20/20 foresight." *CFO: The Magazine for Senior Financial Executives*. CFO Publishing, Boston, MA.

Block, M. R. (March 1994). "ISO/TC207: Developing an international environmental management standard." *The European Marketing Guide*. SIM-COM, Inc., Atlanta, GA.

Bobbit C. E., Jr. (October 1993). "Conduct more effective audits." *Quality Press*. Milwaukee, WI.

Byrnes, D. J. (November/December 1993). "ISO 9000-style eco-audits." *PI Quality*. Hitchcock Publishing Co., Carol Stream, IL.

Chauvel, A-M. (January 1994). "Quality in Europe: Toward the year 2000." *Quality Management Journal*. ASQC, Milwaukee, WI.

Cook, N. P. (July/August 1993). "Quality system, poor products?" *ISO 9000 News: The International Journal of the ISO 9000 Forum*. ISO Central Secretariat, Geneva, Switzerland.

Durant, A. C., and Durant, I. (October 1993). "The role of ISO 9000 standards in continuous improvement." *Quality Systems Update*. CEEM Information Services, Fairfax, VA.

Durant, I., McRobert, C., Middleton, D., and Tirato, J. (May 1992). *Quality Systems Update*. CEEM Information Systems, Fairfax, VA.

Dzus, G. (November 1991). "Planning a successful ISO 9900 assessment." *Quality Progress*.

Dzus, G., and Sykes, E. G., Sr. (October 1993). "How to survive ISO 9000 surveillance." *Quality Progress*. Milwaukee, WI.

Eade, T., and Byrnes, D. J. (September/October 1993). "Documentation per ISO 9000." *PI Quality*. Hitchcock Publishing Co., Carol Stream, IL.

Earnshaw, D. (November 1993). "The EC's eco-management and audit scheme." *The EC Marketing Guide*.

Edelstein, D. V. (November 1993). "Software." *The EC Report on Industry*.

Egan, L. (November 1993). "Software configuration management." *Continuous Improvement*. National ISO 9000 Support Group, Alto, MI.

Finlay, J. S. (August 1992). "ISO 9000, Malcolm Baldrige award guidelines and Deming/SPC-based TQM—A comparison." *Quality Systems Update*. CEEM Information Systems, Fairfax, VA.

Garavaglia, P. L. (October 1993). "How to insure transfer of training." *Training and Development*. American Society for Training and Development, Alexandria, VA.

Gladis, S. D. (July 1993). "Are you the write type?" *Training and Development*. American Society for Training and Development, Alexandria, VA.

Goult, R. (January-May 1992). "ISO implementing an ISO 9000 series." *Quality Systems Update*. CEEM Information Services, Fairfax, VA.

Grounds, R. (October 1993). "Employee involvement: A major change in direction." *Quality Digest*. QCI International, Red Bluff, CA.

Guzzetta, S. (September/October 1993). "How ISO 9000 changed supplier quality assurance." *PI Quality*. Hitchcock Publishing Co., Carol Stream, IL.

Hanrahan, J. J., and Baltus, T. A. (May 15–16). "Efficient engineering through computer-aided design of experiments." *Proceedings of the 41st Annual International Appliance Technical Conference*.

Harral, W. M., and Berg, D. L. (Fall 1993). "Implementing TQM in an ISO framework." *ASQC Automotive Division Newsletter*. Detroit, MI.

Hayes, R. H., and Wheelwright, S. C. (1984). *Restoring Our Competitive Edge: Competing Through Manufacturing*. Wiley, New York.

Hovermale, R. A. (February 1994). "ISO 9000—Continual improvement." *The European Marketing Guide*. SIMCOM, Atlanta, GA.

Howe, K. R., and Dougherty, K. C. (December 1993). "Ethics, institutional review boards, and the changing face of educational research." *Educational Researcher*. AERA.

Jeffrey, N. (January 1994). "Waste not . . . Minimizing and disposing of hazardous waste requires more than lip service." *American Printer*. Maclean Hunter Publishing Co., Chicago.

Kinni, T. B. (October 1993). "Preparing for fast-track ISO 9000 registration." *Quality Digest*. QCI International, Red Bluff, CA.

Kinni, T. B. (January 1994). "Reengineering primer." *Quality Digest*. QCI International, Red Bluff, CA.

Kochan, A. (October 1993). "ISO 9000: Creating a global standardization process." *Quality*. Hitchcock Publishing Co., Carol Stream, IL.

Kolka, J. D. (November 1993). "ISO 9000 and EC medical devices." *The EC Report on Industry*. Single Internal Market Communications, Atlanta, GA.

Kolka, J. D. (March 1994). "Product design, product software and product liability." *The European Marketing Guide*. SIMCOM, Atlanta, GA.

Kromrey, J. D. (May 1993). Ethics and data analysis." *Educational Researcher*. AERA.

Landrum, R. (December 1993). "12 reasons to implement ISO 9000." *Quality Digest*. QCI International, Red Bluff, CA.

LeDoux, T. J. (December 1993/January 1994). "ISO 9000: What you don't know might hurt you!" *Continuous Journey*. American Productivity and Quality Center, Houston, TX.

Mar, W. (October 1993). "Seven keys to better project teams." *Quality Digest.* QCI International, Red Bluff, CA.

Marash, I. R. (February 1994). "ISO 9000 and the medical device GMP." *The European Report on Industry.*

McIntosh, B. (November 1993). "European Community legislation on machinery." *The EC Report on Industry.* Single Internal Market Communications, Atlanta, GA.

Paton, S. M. (August 1991). "The Baldrige, the Deming, ISO 9000 and You." *Quality Digest.*

Petrick, K. (February 1994). "ISO 9000 and the environment—Competing interests." *The European Report on Industry.*

Rabbit, J. T., Bergh, P. A., and Dror, Y. (April 1993). "Capture a quality image with ISO 9000." *INTECH Applying Technology.*

Raymond, C. (November 1993). "An afternoon with an admiral." *OEM Off-Highway.* Johnson Hill Press, Fort Atkison, WI.

Silver, C. H., Jr. (November 1993). "Is your company being torn apart by teamwork?" *OEM Off-Highway.* Johnson Hill Press, Fort Atkison, WI.

Skrabec, Q. R., Jr. (January 1994). "Integrating quality control into your TQM process." *Quality Digest.* QCI International, Red Bluff, CA.

Stamatis, D. H., Epstein, I., and Cooney, R. P. (June 1993). "Documenting personnel qualifications." *Quality Systems Update.* CEEM Information Services, Fairfax, VA.

Stamatis, D. H. (September 1993). "FMEA fulfills prevention intent of ISO 9000." *Quality Systems Update.* CEEM Information System, Fairfax, VA.

Stamatis, D. H. (November 1993). "ISO implementation: A systematic approach." *Proceedings: 1st International Conference on ISO 9000.* Lake Buena Vista, FL.

Stout, G. (October 1993). "Quality practices in Europe." *Quality.* Hitchcock Publishing Co., Carol Stream, IL.

Tartikoff, J. (November 1993). "An environmental perspective on ISO 9000." *Proceedings: 1st Annual International Conference on ISO 9000.* Lake Buena Vista, FL.

Teich, A. H. (1993). *Technology and the Future.* St. Martin's Press, New York.

Wayman, W. R. (January 1994). "ISO 9001: A guide to effective design reviews." *Quality Digest.* QCI International, Red Bluff, CA.

Wolak, J. (March 1994). "ISO 9000—A software market." *Quality.* Hitchcock Publishing, Carol Stream, IL.

Index

For Product Safety Concerns and Information please contact our EU
representative GPSR@taylorandfrancis.com Taylor & Francis Verlag GmbH,
Kaufingerstraße 24, 80331 München, Germany

Printed and bound by CPI Group (UK) Ltd, Croydon, CR0 4YY
08/05/2025
01864538-0002